MATHEMATICS
SIMPLIFIED & SELF-TAUGHT

UNIT 1
ARITHMETIC & COMPUTATIONAL SKILLS

DAVID FRIEDER

A MATHWORKS PUBLICATION
SAN DIEGO
WWW.MATHWORKS-TRAINING.COM

Copyright © 2012 by David Frieder

All rights reserved. No part of this book may be reproduced by any means without permission in writing from the author, except by a reviewer who wishes to quote brief excerpts in connection with a review.

ISBN-13: 978-1478126263

*To Tilay Angbetic,
my partner, soulmate, and
love of my life.*

About The Author

David Frieder has been teaching math to students and adults since 1970 at schools including New York University, Hofstra University, Staten Island College, and LaGuardia Community College. He is the author of *SAT Math Workbook*, *GMAT/GRE Math Review*, *Algebra Simplified and Self-Taught*, and *Clear and Simple Geometry*. In 1965, he received his B.S. in Mathematics from the Polytechnic Institute of New York, and in 1966, his M.A. in Mathematics from the University of Michigan.

In 1987, he co-founded Mathworks, the Math School for Adults, in New York City. Since then, the school has successfully helped thousands of students, adults, and employees learn and refresh all the math skills they studied in school, including those found on the high school equivalency exam, college and graduate school entrance exams, and job application and promotion exams.

Reader Reviews of
GRE/GMAT Math Review by David Frieder

"I'm a self-proclaimed "Math Stupid." It has been nearly 5 years since my last math course and I have forgotten it all. I purchased this book in order to "relearn all my math." I gave myself roughly a month to steadily work through the book's sections and practice questions...and walked away with a great understanding and ability. I even learned things that didn't make sense to me throughout high school and college."

- *Lassie VeUss (Washington, DC)*

"I knew my math skills were weak and I was looking for a review book that would start from the basics and work its way into the more challenging material. Included in this book are tons of practice problems to help you gain the skills you need. I thank the author for producing such a comprehensive review book."

- *Kurt M. Bertram (Boston, MA)*

"I am writing this after trying different books and methods for teaching math. What I have seen and experienced is that this book breaks the concepts down into lucid and simple terms and walks the person through each step. It has been a very good tool. My students were able to understand the concepts when they just looked at how they were used to solve problems. It is good for beginners and for people who have relatively less background in math."

- *Deepak Madabushi*

"I always struggled with math. I found a lot of this was due to the fact that incompetent individuals taught and only showed you the text book examples which only managed to show the easiest questions. I learned with this book, though, that it is the material and how it is presented. The author is very helpful and everything is broken down so you can observe multiple problems and then practice them. At the end of each section you can practice what you learned to ensure that you have the hang of it. The lessons make sense, steps are generally not skipped. In conclusion, after spending a few hours researching texts out there and about 20 hours working through it question by question, I have no reservations in recommending this book to those that have any questions/concerns or desire to ensure your math is bulletproof on any exam. It teaches you from the basics (most math is based on simple arithmetic and algebra). I may give it to relatives to help with their math in HS/college/university when I am done as it makes more sense than most HS texts."

- *J. Brown (Cleveland, Ohio)*

Introduction

The purpose of the *Mathematics–Simplified & Self-Taught Series* is to provide the reader with a total review of all the math skills they learned in school. It also serves as a perfect preparation for students and adults who are getting ready to take exams, such as job application exams, Civil Service exams, the High School Equivalency Test (GED), the Scholastic Aptitude Test (SAT), the ACT, the Graduate Management Admission Test (GMAT), and the Graduate Record Exam (GRE).

Each unit in the series is written as a self-teaching guide and contains a large number of Sample Problems which illustrate the principles and procedures shown in the text. The solutions show step-by-step explanations of how to proceed. The Sample Problems are immediately followed by similar Practice Problems for the reader to try on his/her own. In addition, a Review Test of twenty-five questions is given at the end of each unit.

This first unit, *Arithmetic & Computational Skills*, covers whole numbers, fractions, proportions, decimals, and percents. An appendix is also included which shows how to use the various Clear keys and Memory keys on a basic calculator.

CONTENTS

Chapter 1 – Whole Numbers
- 1.1 Whole Numbers — 1
- 1.2 Rounding Off Whole Numbers — 1
- 1.3 Fundamental Operations — 2
- 1.4 Prime Numbers And Composite Numbers — 2

Chapter 2 – Fractions
- 2.1 Fractions — 4
- 2.2 Types Of Fractions — 4
- 2.3 Raising Fractions To Higher Terms — 5
- 2.4 Reducing Fractions To Lower Terms — 6
- 2.5 Adding And Subtracting Like Fractions — 8
- 2.6 Adding And Subtracting Unlike Fractions — 10
- 2.7 Multiplying Fractions — 12
- 2.8 Dividing Fractions — 13
- 2.9 Complex Fractions — 14
- 2.10 Comparing Fractions — 15
- 2.11 Finding The Missing Term Of A Proportion — 16

Chapter 3 – Decimals
- 3.1 Decimals — 18
- 3.2 Rounding Off Decimals — 19
- 3.3 Adding And Subtracting Decimals — 19
- 3.4 Multiplying Decimals — 20
- 3.5 Dividing A Decimal By A Whole Number — 21
- 3.6 Dividing A Decimal By A Decimal — 23
- 3.7 Comparing Decimals — 25
- 3.8 Changing Decimals To Fractions — 26
- 3.9 Changing Fractions To Decimals — 27

Chapter 4 – Percents
- 4.1 Percents — 28
- 4.2 Changing Percents To Decimals — 28
- 4.3 Changing Percents To Fractions — 30

Arithmetic Review Test — 35

Solutions To Arithmetic Practice Problems — 39

Solutions To Arithmetic Review Test — 46

Appendix
- A.1 A Tour Of The Basic Calculator — 49
- A.2 Calculator Keystroke Notation — 50
- A.3 The Clear Keys — 50
- A.4 The Memory Keys — 51

CHAPTER 1

1.1 WHOLE NUMBERS

The set of *whole numbers* is an infinite set of numbers consisting of all the counting numbers, 1, 2, 3, etc., and the number 0.

Whole Numbers
{0, 1, 2, 3, ...}

The value of each digit in a whole number is determined by the particular *place* it occupies in the number. For example, in the number 357, the 3 is in the hundreds place, and thus represents the value 300; the 5 is in the tens place, and thus represents the value 50; and the 7 is in the ones place, and thus represents the value 7. A summary of whole number *place-values* is given below.

$357 = 300 + 50 + 7$

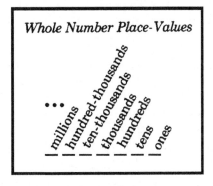

1.2 ROUNDING OFF WHOLE NUMBERS

When only an approximate value of a whole number is needed, we use the following procedure to *round off* the number to a particular place. Note that in the example in the margin, the symbol ≈ means "is approximately equal to."

Round off 9371 to the nearest **hundred**

> ***To Round Off a Whole Number to a Particular Place:***
>
> 1. Identify the digit in the place being rounded off.
> 2. If the digit to its right is less than 5, leave the identified digit as it is.
>   If the digit to its right is 5 or more, increase the identified digit by 1.
> 3. Replace all the digits to the right of the identified digit by zeros.

9[3]71
↓
9[3]71
9[4]71
9400
9371 ≈ 9400

Rounding off whole numbers is particularly useful in estimating the results of arithmetic operations. To do this, we usually round off the left most digit of each number in the operation, and replace all the other digits by zeros. For example, to estimate the value of 47×624, we round off 47 to 50 and 624 to 600 and then multiply. As shown in the margin, this gives us an estimated value of 30,000.

$47 \times 624 \approx 50 \times 600$
$\approx 30,000$

Page 1

1.2 PRACTICE PROBLEMS

(Solutions to all Practice Problems begin on Page 39)

Round off the following whole numbers to the place indicated.

1. 34,682 to the nearest thousand
2. 5,416,248 to the nearest hundred-thousand
3. 299,961 to the nearest hundred
4. 68,199 to the nearest ten-thousand

1.3 FUNDAMENTAL OPERATIONS

Each of the fundamental operations of arithmetic–addition, subtraction, multiplication, and division–has associated with it a special set of terms that identify the numbers used in the operation as well as the result of the operation.

> **Terms Associated with Operations**
>
> Addition: ***Addend + Addend = Sum***
> Subtraction: ***Minuend − Subtrahend = Difference***
> Multiplication: ***Factor × Factor = Product***
> Division: ***Dividend ÷ Divisor = Quotient***
>
> $$\text{Divisor} \overline{\smash{)}\text{Dividend}}^{\text{Quotient}}$$

$3 + 5 = 5 + 3$
$7 \times 9 = 9 \times 7$

$6 - 2 \neq 2 - 6$
$8 \div 4 \neq 4 \div 8$

Remember that only addition and multiplication have the property that when we reverse the order of the numbers, we get the same result. This property, called the ***commutative property*** of addition and multiplication, is not true for subtraction or division. In the examples shown in the margin, the symbol $\neq$ means "is not equal to."

1.4 PRIME NUMBERS AND COMPOSITE NUMBERS

A whole number, M, is said to be ***divisible*** by a whole number, N, if N divides into M without a remainder ("evenly"). For example, since 4 divides into 20 without a remainder, we say that 20 is divisible by 4.

When a whole number, M, is divisible by a whole number, N, then N is said to be a ***divisor*** or ***factor*** of M, and M is said to be a ***multiple*** of N. In the previous example, 4 is a divisor of 20, and 20 is a multiple of 4.

Each whole number greater than 0 has an infinite number of multiples, but only a limited number of divisors. For example, the multiples of 6 are 6, 12, 18, 24, 30, etc., while the divisors of 6 are only 1, 2, 3, and 6.

A whole number that has only two divisors–the number itself and the number 1–is called a ***prime number***. A whole number that has more than two divisors is called a ***composite number***. Examples of both types of numbers are shown in the margin. Note that 0 and 1 are not considered prime numbers or composite numbers. Also, note that the only even prime number is 2. All even numbers greater than 2 have 2 as a divisor and are therefore composite numbers.

Every composite number can be written as a product of prime numbers. One way we can do this is described below.

Prime Numbers
2, 3, 5, 7, 11, 13, 17, ...

Composite Numbers
4, 6, 8, 9, 10, 12, 14, ...

To Obtain the Prime Factorization of a Number:

1. Find any pair of numbers whose product is the given number, and write the pair at the end of two branches leading from the number.
2. Continue this branching process until every number at the end of a branch is a prime number.
3. Form a product of all the prime numbers at the ends of the branches.

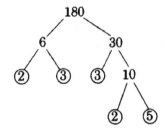

$180 = 2 \times 3 \times 3 \times 2 \times 5$

1.4 PRACTICE PROBLEMS

Write each of the following numbers as a product of prime number factors.

1. 60   2. 315   3. 176   4. 825

CHAPTER 2

2.1 FRACTIONS

$$\frac{\text{numerator}}{\text{denominator}}$$

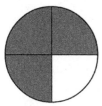

$$\frac{\text{part}}{\text{whole}} = \frac{3}{4}$$

$$7 \div 3 = 3\overline{)7} = \frac{7}{3}$$

$$\frac{\text{cents in a quarter}}{\text{cents in a dime}} = \frac{25}{10}$$

Fractions consist of two numbers separated by a line called a *fraction bar.* The number above the line is called the **numerator** of the fraction, and the number below the line is called the **denominator**. Depending upon the way a fraction is used, these numbers have several different meanings.

The most common way of using a fraction is to represent a *part of a whole*. For example, the fraction $\frac{3}{4}$ represents 3 of the 4 equal parts into which the whole circle in the margin has been divided.

Another way of using a fraction is to indicate **division** of one number by another. For example, the fraction $\frac{7}{3}$ can be used to indicate the division of 7 by 3 (or equivalently, 3 divided into 7).

A third way of using a fraction is to express a *ratio*, or comparison, between two quantities. For example, the fraction $\frac{25}{10}$ can used to express the ratio of the number of cents in a quarter to the number of cents in a dime.

These different interpretations of a fraction are summarized below.

> *Interpretations of a Fraction*
>
> 1. A part of a whole: $\dfrac{\text{part}}{\text{whole}}$
>
> 2. Division of one number by another: $\dfrac{\text{dividend}}{\text{divisor}}$
>
> 3. Ratio of two quantities: $\dfrac{\text{quantity A}}{\text{quantity B}}$

2.2 TYPES OF FRACTIONS

Proper Fractions

$$\frac{3}{4}, \frac{5}{9}, \frac{1}{2}$$

Improper Fractions

$$\frac{5}{2}, \frac{7}{7}, \frac{9}{4}$$

Fractions are classified into two categories according to the relative size of their numerators and denominators: (1) *proper fractions*–fractions whose numerators are less than their denominators, and (2) *improper fraction*–fractions whose numerators are either equal to or greater than their denominators. The value of a proper fraction is always less than 1 (one whole), and the value of an improper fraction is either equal to or greater than 1. Any whole number can be put into the form of an improper fraction by placing it over a denominator of 1. For example $8 = \frac{8}{1}$, and $12 = \frac{12}{1}$.

The sum of a whole number and a proper fraction is called a *mixed number*. Mixed numbers are denoted by placing the whole number and the fraction side by side without using the addition symbol. For example, $4 + \frac{8}{9} = 4\frac{8}{9}$.

Mixed Numbers

$4\frac{8}{9}, 5\frac{2}{3}, 8\frac{6}{7}$

A mixed number can be changed into the form of an improper fraction by the following procedure.

To Change a Mixed Number into an Improper Fraction:

1. Multiply the denominator of the fraction by the whole number.
2. Add the resulting product to the numerator of the fraction and place the sum over the original denominator.

$2\frac{5}{7} = \frac{14+5}{7}$

$2\frac{5}{7} = \frac{19}{7}$

By interpreting a fraction as a quotient, we can reverse this procedure and change the improper fraction back into the form of a mixed number.

To Change an Improper Fraction into a Mixed Number:

1. Divide the denominator into the numerator.
2. Express any remainder as a fraction by placing it over the original denominator.

$\frac{19}{7} = 7\overline{)19}$ with quotient 2, remainder 5

$\frac{19}{7} = 2\frac{5}{7}$

2.2 PRACTICE PROBLEMS

Change the following mixed numbers to improper fractions.

1. $5\frac{3}{4}$ 2. $7\frac{3}{8}$ 3. $6\frac{2}{3}$ 4. $12\frac{1}{2}$

Change the following improper fractions to mixed numbers.

5. $\frac{14}{3}$ 6. $\frac{27}{2}$ 7. $\frac{48}{4}$ 8. $\frac{6}{5}$

2.3 RAISING FRACTIONS TO HIGHER TERMS

Fractions that have the same value are said to be *equivalent*. For example, $\frac{2}{4}, \frac{3}{6}$, and $\frac{4}{8}$ are equivalent fractions, all having the same value, $\frac{1}{2}$.

Fractions can be changed into equivalent fractions by multiplying or dividing both their numerators and denominators by the same non-zero number.

When both the numerator and denominator of a fraction are multiplied by the same number greater than 1, the fraction is said to be *raised to higher terms*. Remember that although the numerator and denominator of the new fraction are larger than those of the original fraction, the fractions are equivalent in value.

$$\frac{a}{b} = \frac{a \times c}{b \times c}, c > 1$$

If the denominator of the fraction in higher terms is specified in advance, we can find the corresponding numerator by the following procedure.

$$\frac{3}{5} = \frac{?}{20}$$

$$\frac{3}{5} \underset{4}{\overset{4}{\rightrightarrows}} \frac{12}{20}$$

To Raise a Fraction to Higher Terms:

1. Divide the original denominator into the new specified denominator.
2. Multiply the result by the original numerator.

2.3 PRACTICE PROBLEMS

Change each of the following fractions to an equivalent fraction having the denominator indicated:

1. $\dfrac{3}{4} = \dfrac{?}{20}$ 2. $\dfrac{5}{9} = \dfrac{?}{72}$ 3. $\dfrac{4}{11} = \dfrac{?}{66}$ 4. $\dfrac{8}{3} = \dfrac{?}{12}$

2.4 REDUCING FRACTIONS TO LOWER TERMS

$$\frac{a}{b} = \frac{a \div c}{b \div c}, c > 1$$

$$\frac{6}{8} = \frac{6 \div 2}{8 \div 2} = \frac{3}{4}$$

When both the numerator and denominator of a fraction are divided by the same number greater than 1 (a *common divisor* or *factor*), the fraction is said to be *reduced to lower terms*. Remember that the new fraction is not smaller in value than the original fraction, but has the same value, expressed in smaller ("reduced") terms. Also, fractions, such as $\dfrac{3}{7}$, which cannot be reduced further, are said to be *reduced to lowest terms*.

To reduce a fraction to lower terms, we must find a number which can divide "evenly" (with no remainder) into both the numerator and denominator. With the divisibility tests below, we can quickly check if the numerator and denominator are both divisible by certain common numbers.

Divisibility Tests for 2, 3, 5, and 10

A number is divisible by:

2, if its last digit is even – 0, 2, 4, 6, or 8
3, if the sum of its digits is a number divisible by 3
5, if its last digit is 0 or 5
10, if its last digit is 0

These tests are illustrated in the examples below.

Reduce by 2	Reduce by 3	Reduce by 5	Reduce by 10
$\dfrac{14 \div 2}{38 \div 2} = \dfrac{7}{19}$	$(5+1=6) \downarrow$ $\dfrac{51 \div 3}{72 \div 3} = \dfrac{17}{24}$ $\uparrow (7+2=9)$	$\dfrac{15 \div 5}{40 \div 5} = \dfrac{3}{8}$	$\dfrac{20 \div 10}{30 \div 10} = \dfrac{2}{3}$

Although there are no quick divisibility tests for numbers like 7, 11, and 13, we must remember to try them as well. For example, $\dfrac{21}{28}$ can be reduced to $\dfrac{3}{4}$ by dividing both the numerator and denominator by 7.

SAMPLE PROBLEM

Problem 1: Reduce $\dfrac{1740}{3960}$ to lowest terms.

Explanation and Procedure	Math Steps
1. Since the numerator and denominator both end in 0, divide them by 10.	$\dfrac{1740}{3960} = \dfrac{1740 \div 10}{3960 \div 10}$ $= \dfrac{174}{396}$
2. Since the numerator and denominator both end in an even digit, divide them by 2.	$= \dfrac{174 \div 2}{396 \div 2}$ $= \dfrac{87}{198}$
3. The sum of the digits in the numerator is 15, which is divisible by 3, and the sum of the digits in the denominator is 18, which is also divisible by 3. Therefore, divide the numerator and denominator by 3. Since the result cannot be reduced further, it is reduced to lowest terms.	$= \dfrac{87 \div 3}{198 \div 3}$ $= \dfrac{29}{66}$

2.4 PRACTICE PROBLEMS

Reduce the following fractions to lowest terms.

1. $\dfrac{24}{72}$
2. $\dfrac{135}{243}$
3. $\dfrac{750}{1250}$
4. $\dfrac{420}{560}$

2.5 ADDING AND SUBTRACTING LIKE FRACTIONS

$$\frac{a}{c}+\frac{b}{c}=\frac{a+b}{c}$$
$$\frac{a}{c}-\frac{b}{c}=\frac{a-b}{c}$$

To add or subtract fractions having the same denominators (*like fractions*), we add or subtract their numerators, and place the result over their common denominator. We often **simplify** the result by changing improper fractions to mixed numbers, and by reducing proper fractions to lower terms.

$$\frac{7}{8}+\frac{6}{8}-\frac{1}{8}=\frac{7+6-1}{8}$$
$$=\frac{12}{8}$$
$$=1\frac{4}{8}$$
$$=1\frac{1}{2}$$

To add or subtract mixed numbers, the whole numbers and fractions are treated separately. This is facilitated by arranging the problem vertically, with the whole numbers and fractions lined up. Remember that the final result can often be simplified to lower terms. In the examples below, note that in the one on the left, the result, $14\frac{16}{12}$, is simplified by changing $\frac{16}{12}$ to $1\frac{4}{12}$, and then combining it with 14 to get $15\frac{4}{12}$, or $15\frac{1}{3}$.

$$\begin{array}{r}8\frac{5}{12}\\+6\frac{11}{12}\\\hline 14\frac{16}{12}=15\frac{4}{12}\\=15\frac{1}{3}\end{array} \qquad \begin{array}{r}9\frac{3}{4}\\-2\frac{1}{4}\\\hline 7\frac{2}{4}=7\frac{1}{2}\end{array}$$

Sometimes, when subtracting mixed numbers, it is necessary to borrow 1 from the whole number of the number on the top after the numbers are arranged vertically. This is illustrated in the sample problems that follow.

SAMPLE PROBLEM

Problem 2: Subtract $6 - 2\frac{3}{7}$.

Explanation and Procedure	Math Steps
1. Arrange the problem vertically with the whole numbers lined up.	$\begin{array}{r} 6 \\ -2\frac{3}{7} \\ \hline \end{array}$
2. Since there is no fraction in the top number 6, borrow 1 from it, leaving 5, and change it into the equivalent fraction, $1 = \frac{7}{7}$, having the same denominator as the bottom fraction. Subtract.	$\begin{array}{r} 5\,\cancel{6}\,\frac{7}{7} \\ -2\frac{3}{7} \\ \hline 3\frac{4}{7} \end{array}$

SAMPLE PROBLEM

Problem 3: Subtract $8\frac{2}{3} - 3\frac{5}{9}$.

Explanation and Procedure	Math Steps
1. Arrange the problem vertically with the whole numbers and fractions lined up.	$\begin{array}{r} 8\frac{2}{9} \\ -3\frac{5}{9} \\ \hline \end{array}$
2. Since $\frac{5}{9}$ cannot be subtracted from $\frac{2}{9}$, borrow 1 from the whole number 8 and place it next to $\frac{2}{9}$. Change the mixed number, $1\frac{2}{9}$, to the improper fraction, $\frac{11}{9}$, and subtract. Reduce the result to lowest terms.	$\begin{array}{r} 7\,\cancel{8}\,1\frac{2}{9} = 7\frac{11}{9} \\ -3\frac{5}{9} = 3\frac{5}{9} \\ \hline 4\frac{6}{9} = 4\frac{2}{3} \end{array}$

2.5 PRACTICE PROBLEMS

Perform the following additions and subtractions. Reduce all answers to lowest terms.

1. $5\frac{3}{4} + 2\frac{3}{4}$ 2. $12\frac{5}{9} - 7\frac{2}{9}$ 3. $8 - 5\frac{5}{6}$ 4. $16\frac{4}{7} - 9\frac{6}{7}$

2.6 ADDING AND SUBTRACTING UNLIKE FRACTIONS

To add or subtract fractions having different denominators (*unlike fractions*), we first change them to equivalent fractions having a common denominator (equivalent like fractions), and then proceed as in the previous section.

For example, to add $\frac{3}{4}+\frac{5}{6}$, in which the fractions have different denominators, we first change them to equivalent fractions having a common denominator. In general, a **common denominator** for a given set of fractions is any number that is divisible by all the denominators, or in other words, a common multiple of the denominators. One such number will always be the product of the denominators. Thus, in this example, we can use 24, the product of 4 and 6. We can also use any of the other common multiples indicated below. Note that 12, the smallest common multiple, is usually referred to as the *least common multiple* of the denominators, or the *least common denominator (L.C.D.)*.

Common Denominators of $\frac{3}{4}$ and $\frac{5}{6}$

Multiples of 4 : 4, 8, $\boxed{12}$, 16, 20, $\boxed{24}$, 28, 32, $\boxed{36}$, ...

Multiples of 6 : 6, $\boxed{12}$, 18, $\boxed{24}$, 30, $\boxed{36}$, ...

After choosing one of these denominators (the choice is arbitrary), by using the procedure for changing the fractions to higher terms, we change each of the fractions to an equivalent fraction having this denominator. Then we add the resulting numerators. For example, if we choose the common denominator 12, we get:

$$\frac{3}{4}=\frac{9}{12}$$
$$+\frac{5}{6}=\frac{10}{12}$$
$$\frac{19}{12}=1\frac{7}{12}$$

> ### *To Add or Subtract Unlike Fractions:*
> 1. Find a common denominator for the fractions (a common multiple of the denominators). Remember that one can always be obtained by multiplying the denominators.
> 2. By using the procedure for changing fractions to higher terms, change each of the fractions to the common denominator.
> 3. Add or subtract the numerators and place the result over the common denominator.
> 4. Simplify the resulting fraction.

$$\frac{5}{6} = \frac{40}{48}$$
$$-\frac{3}{8} = \frac{18}{48}$$
$$\frac{22}{48} = \frac{11}{24}$$

SAMPLE PROBLEM

Problem 4: Add $4\frac{3}{5} + 5\frac{2}{7} + 3\frac{1}{2}$.

Explanation and Procedure	Math Steps
1. Arrange the numbers vertically with the whole numbers and fractions lined up. Change each of the fractions to the common denominator 70, the product of the denominators. Add the whole numbers and fractions separately, and simplify the result.	$4\frac{3}{5} = 4\frac{42}{70}$ $5\frac{2}{7} = 5\frac{20}{70}$ $+3\frac{1}{2} = 3\frac{35}{70}$ $12\frac{97}{70} = 13\frac{27}{70}$

SAMPLE PROBLEM

Problem 5: Subtract $9\frac{3}{7} - 3\frac{1}{2}$.

Explanation and Procedure	Math Steps
1. Arrange the numbers vertically with the whole numbers and fractions lined up. Change each of the fractions to the common denominator 14.	$9\frac{3}{7} = 9\frac{6}{14}$ $-3\frac{1}{2} = 3\frac{7}{14}$
2. Since we cannot subtract $\frac{7}{14}$ from $\frac{6}{14}$, borrow 1 from 9 and place it next to $\frac{6}{14}$. Change the mixed number $1\frac{6}{14}$ to the improper fraction $\frac{20}{14}$, and subtract.	$9\frac{3}{7} = {}^8\cancel{9}1\frac{6}{14} = 8\frac{20}{14}$ $-3\frac{1}{2} = \quad 3\frac{7}{14} = 3\frac{7}{14}$ $\qquad\qquad\qquad 5\frac{13}{14}$

2.6 PRACTICE PROBLEMS

Perform the following additions and subtractions. Reduce all answers to lowest terms.

1. $6\frac{3}{5} + 2\frac{5}{7}$ 2. $2\frac{2}{3} + 4\frac{1}{4} + 1\frac{3}{5}$ 3. $8\frac{2}{3} - 3\frac{1}{2}$ 4. $19\frac{2}{5} - 12\frac{3}{4}$

2.7 MULTIPLYING FRACTIONS

Multiplying fractions is usually indicated by the word "of." For example, $\frac{3}{4}$ of $\frac{5}{7}$ means $\frac{3}{4} \times \frac{5}{7}$.

Unlike when we are adding or subtracting fractions, it is not necessary to change the fractions to a common denominator. Instead, we simply multiply the numerators by the numerators, and the denominators by the denominators.

$$\frac{a}{b} \times \frac{c}{d} = \frac{a \times c}{b \times d}$$

$$\frac{3}{4} \times \frac{5}{7} = \frac{3 \times 5}{4 \times 7} = \frac{15}{28}$$

If any of the numbers in the product are whole numbers or mixed numbers, they should be changed to improper fractions before multiplying. Remember that a whole number can be changed to an improper fraction by placing it over a denominator of 1.

$$\frac{2}{9} \times 5 \times 4\frac{2}{3} = \frac{2}{9} \times \frac{5}{1} \times \frac{14}{3}$$
$$= \frac{2 \times 5 \times 14}{9 \times 1 \times 3}$$
$$= \frac{140}{27}$$
$$= 5\frac{5}{27}$$

Sometimes we can simplify a product of fractions before multiplying them by "canceling" common factors in the numerators and denominators.

For example, when multiplying $\frac{5}{6} \times \frac{9}{20}$, we can divide out ("cancel") a common factor of 5 in 5 and 20, and a common factor of 3 in 9 and 6. (Notice that the numerator and denominator do not have to be part of the same fraction.) Then, we multiply the remaining factors and get the result $\frac{3}{8}$. This result is the same as the one we would get by multiplying first and reducing after.

$$\begin{array}{c|c}
\textit{Cancel First} & \textit{Multiply First} \\
\dfrac{5}{6} \times \dfrac{9}{20} = \dfrac{\overset{1}{\cancel{5}}}{\underset{2}{\cancel{6}}} \times \dfrac{\overset{3}{\cancel{9}}}{\underset{4}{\cancel{20}}} & \dfrac{5}{6} \times \dfrac{9}{20} = \dfrac{45}{120} \\
= \dfrac{1}{2} \times \dfrac{3}{4} & = \dfrac{45 \div 5}{120 \div 5} \\
 & = \dfrac{9 \div 3}{24 \div 3} \\
= \dfrac{3}{8} & = \dfrac{3}{8}
\end{array}$$

It is usually much easier to find common factors to cancel before multiplying than it is to find common factors to reduce after multiplying.

SAMPLE PROBLEM

Problem 6: Multiply $\dfrac{4}{5} \times 15 \times 1\dfrac{3}{8}$.

Explanation and Procedure	Math Steps
1. Change the whole number and mixed number into improper fractions.	$\dfrac{4}{5} \times 15 \times 1\dfrac{3}{8}$ $= \dfrac{4}{5} \times \dfrac{15}{1} \times \dfrac{11}{8}$
2. Cancel the common factors in the numerators and denominators, and multiply the remaining factors.	$= \dfrac{\overset{1}{\cancel{4}}}{\underset{1}{\cancel{5}}} \times \dfrac{\overset{3}{\cancel{15}}}{1} \times \dfrac{11}{\underset{2}{\cancel{8}}}$ $= \dfrac{33}{2}$
3. Simplify the final result.	$= 16\dfrac{1}{2}$

2.7 PRACTICE PROBLEMS

Perform the following multiplications. Reduce all answers to lowest terms.

1. $\dfrac{2}{3} \times \dfrac{6}{7} \times \dfrac{3}{4}$ 2. $2\dfrac{2}{3} \times 1\dfrac{4}{5}$ 3. $5 \times \dfrac{2}{3} \times 1\dfrac{1}{5}$ 4. $1\dfrac{4}{5} \times 2\dfrac{2}{3} \times 2\dfrac{1}{2}$

2.8 DIVIDING FRACTIONS

To divide fractions, we first invert the divisor (turn it upside down), and then multiply the resulting fractions. Remember that when using the symbol ÷ ("divided by"), the divisor is the fraction on the right.

$$\dfrac{a}{b} \div \dfrac{c}{d} = \dfrac{a}{b} \times \dfrac{d}{c}$$

$$\dfrac{3}{4} \div \dfrac{7}{8} = \dfrac{3}{\underset{1}{\cancel{4}}} \times \dfrac{\overset{2}{\cancel{8}}}{7}$$

$$= \dfrac{6}{7}$$

As before, we can cancel common factors in the numerators and denominators, but only *after* we invert the divisor and change the operation to multiplication.

If the divisor or dividend is a whole number or a mixed number, it should first be changed to an improper fraction before proceeding.

SAMPLE PROBLEM

Problem 7: Divide $3\frac{1}{2} \div 2\frac{5}{8}$.

Explanation and Procedure	Math Steps
1. Change the mixed numbers to improper fractions.	$3\frac{1}{2} \div 2\frac{5}{8} = \frac{7}{2} \div \frac{21}{8}$
2. Invert the divisor and change the division to multiplication.	$= \frac{7}{2} \times \frac{8}{21}$
3. Cancel the common factors in the numerators and denominators, and multiply the remaining factors.	$= \frac{\cancel{7}^1}{\cancel{2}_1} \times \frac{\cancel{8}^4}{\cancel{21}_3}$ $= \frac{4}{3}$
4. Simplify the final result.	$= 1\frac{1}{3}$

2.8 PRACTICE PROBLEMS

Perform the following divisions. Reduce all answers to lowest terms.

1. $\frac{5}{6} \div \frac{4}{7}$
2. $8 \div \frac{4}{9}$
3. $2\frac{2}{3} \div 1\frac{7}{9}$
4. $2\frac{2}{5} \div 6$

2.9 COMPLEX FRACTIONS

$\dfrac{5 - \frac{2}{3}}{4}$, $\dfrac{\frac{1}{2} + 6}{\frac{5}{7}}$

A fraction that contains at least one other fraction in its numerator or denominator is called a *complex fraction*. There are two ways of simplifying a complex fraction. The first is to combine the numbers in the numerator and denominator separately, and then divide the resulting numerator by the resulting denominator. For example,

$$\frac{\frac{1}{4} + \frac{1}{3}}{\frac{5}{6}} = \frac{\frac{7}{12}}{\frac{5}{6}}$$

$$= \frac{7}{12} \div \frac{5}{6}$$

$$= \frac{7}{12} \times \frac{6}{5}$$

$$= \frac{42}{60} = \frac{7}{10}$$

Alternatively, we can multiply every term in the complex fraction by a common denominator of the fractions contained within it, leaving a fraction free of other fractions. For example, to simplify the complex fraction in the preceding example, we can multiply every term by 12, a common denominator of $\frac{1}{4}, \frac{1}{3}$, and $\frac{5}{6}$, and get:

$$\frac{\frac{1}{4}+\frac{1}{3}}{\frac{5}{6}} = \frac{\overset{3}{\cancel{12}} \times \frac{1}{\cancel{4}} + \overset{4}{\cancel{12}} \times \frac{1}{\cancel{3}}}{\overset{2}{\cancel{12}} \times \frac{5}{\cancel{6}}}$$

$$= \frac{3+4}{10}$$

$$= \frac{7}{10}$$

2.9 PRACTICE PROBLEMS

Simplify the following complex fractions.

1. $\dfrac{\frac{2}{5}}{\frac{3}{4}}$ 2. $\dfrac{4-\frac{2}{7}}{6}$ 3. $\dfrac{\frac{1}{2}+\frac{2}{3}}{5}$ 4. $\dfrac{\frac{3}{4}+2}{\frac{1}{2}}$

2.10 COMPARING FRACTIONS

There are several methods we can use to compare the values of fractions. One is to change the fractions to a common denominator (as we do in addition and subtraction), and then compare their numerators.

For example, to compare $\frac{4}{7}$ and $\frac{5}{9}$, we can change each of the fractions to the common denominator 63:

$$\frac{4}{7} = \frac{36}{63} \text{ and } \frac{5}{9} = \frac{35}{63}$$

Then, when we compare their numerators, we see that $\frac{36}{63}$ is greater than $\frac{35}{63}$, and therefore conclude that $\frac{4}{7}$ is greater than $\frac{5}{9}$.

Another method is to cross-multiply their numerators and denominators and compare the resulting products:

Page 15

Notice that the cross-products, 36 and 35, correspond to the numerators we compared in the first method. Since 36 is greater than 35, we again conclude that $\frac{4}{7}$ is greater than $\frac{5}{9}$.

In the summary below, the symbol > means "is greater than," and the symbol < means "is less than."

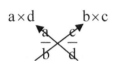

> **To Compare Two Fractions $\frac{a}{b}$ and $\frac{c}{d}$**
>
> 1. Cross-multiply the numerators and denominators, and write the resulting products above the factions.
>
>   If $a \times d > b \times c$, then $\frac{a}{b} > \frac{c}{d}$
>
>   If $a \times d = b \times c$, then $\frac{a}{b} = \frac{c}{d}$
>
>   If $a \times d < b \times c$, then $\frac{a}{b} < \frac{c}{d}$

2.10 PRACTICE PROBLEMS

In each of the following groups of fractions, which fraction is the largest.

1. $\frac{6}{11}, \frac{5}{9}$
2. $\frac{9}{13}, \frac{11}{15}$
3. $\frac{5}{7}, \frac{2}{3}, \frac{3}{4}$

2.11 FINDING THE MISSING TERM OF A PROPORTION

$\frac{\text{Quarter}}{\text{Dime}} = \frac{25}{10}$

$= \frac{25 \div 5}{10 \div 5}$

$= \frac{5}{2}$

Remember, one of the ways to use a fraction is to express a ratio, or comparison, between quantities. For example, the fraction $\frac{25}{10}$ can be used to express the ratio of the number of cents in a quarter to the number of cents in a dime. As with fractions, ratios can be reduced to lower terms. Therefore, the ratio $\frac{25}{10}$ ("25 to 10") is equivalent to the ratio $\frac{5}{2}$ ("5 to 2"). Note that ratios are sometimes written using a colon, such as 5:2.

Ratios can involve more than two terms. For example, if three friends invest $3,000, $6,000, and $12,000 in a stock, respectively, we can say that the invested their money in the ratio of $3,000 to $5,000 to $8,000 or 3:5:8.

When two ratios (fractions) are equal, they form a **proportion**. Thus, the statement $\frac{25}{10} = \frac{5}{2}$ is an example of a proportion. We read this statement as "25 is to 10 as 5 is to 2". Each of the four numbers in a proportion is called a *term* of the proportion. Sometimes we are given only three of the terms and are asked to find the fourth. For example, in the proportion $\frac{3}{5} = \frac{21}{N}$ we are missing the term N. To find N, we use the fact that when two fractions are equal, the cross-products of their numerators and denominators are equal.

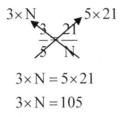
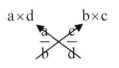

$$3 \times N = 5 \times 21$$
$$3 \times N = 105$$

Since 3 times N equals 105, N must equal 105 divided by 3. Thus, we get:

$$3 \times N = 105$$
$$N = \frac{105}{3}$$
$$N = 35$$

To Find the Missing Term of a Proportion:

1. Cross-multiply the numerators and denominators and set the two products equal.
2. Divide the number in the product containing the missing term into the other product.

2.11 PRACTICE PROBLEMS

Find the missing term, N, of each of the following proportions.

1. $\frac{7}{9} = \frac{N}{72}$
2. $\frac{19}{12} = \frac{76}{N}$
3. $\frac{N}{16} = \frac{15}{20}$
4. $\frac{30}{N} = \frac{45}{27}$

CHAPTER 3

3.1 DECIMALS

Fractions, such as $\frac{3}{10}$, $\frac{17}{100}$, and $\frac{9}{1000}$, whose denominators are powers of 10 (10, 100, 1000, etc.), are called **decimal fractions**, or simply, **decimals**.

$\frac{3}{10} = .3$

$\frac{17}{100} = .17$

$\frac{9}{1000} = .009$

Decimal fractions are usually written in a special shorthand in which the numerator is placed to the right of a dot called a **decimal point**. The denominator is not actually written, but instead is determined in the following way: If the numerator ends one place to the right of the decimal point, the denominator is 10; if the numerator ends two places to the right of the decimal point, the denominator is 100; if the numerator ends three places to the right of the decimal point, the denominator is 1000; and so on. Note that to write $\frac{9}{1000}$ using this shorthand, we must use two zeros as placeholders so that the numerator ends the required three places to the right of the decimal point.

$19\frac{7}{10} = 19.7$

$8\frac{3}{100} = 8.03$

Numbers that consist of a whole number and a decimal are called **mixed decimals**. As shown in the two examples in the margin, mixed decimals are written with the whole number to the left of the decimal point and the decimal fraction to the right of it. The values of the places in a mixed decimal are summarized below. Note that the decimal point does not occupy a place by itself, but simply separates the whole number places from the decimal places.

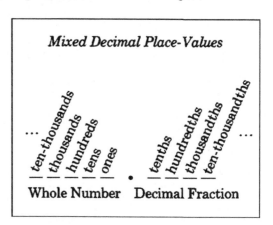

When reading mixed decimals, the decimal point is read as the word "and." This signals the separation between the whole number and the decimal fraction. For example, 347.15 is read as "three hundred forty seven *and* fifteen hundredths."

Remember that depending on where they are placed, zeros have different effects on the value of a decimal. On the one hand, zeros placed *between* the decimal point and the first digit in a decimal make the value of the decimal smaller.

$$.9 = \frac{9}{10}, \quad .09 = \frac{9}{100}, \quad .009 = \frac{9}{1000}$$

On the other hand, zeros placed *to the right of* the last digit in a decimal do not change the value of the fraction at all:

$$.7 = \frac{7}{10}, \quad .70 = \frac{70}{100}, \quad .700 = \frac{700}{1000}$$

After reducing these fractions, we see that each has the same value, $\frac{7}{10}$.

3.2 ROUNDING OFF DECIMALS

The procedure for rounding off decimals is very similar to the procedure for rounding off whole numbers. The only notable difference is that when rounding off decimals, the zeros to the right of the rounded off place have no value and can be dropped.

To Round Off a Decimal:
1. Identify the digit in the place being rounded off.
2. If the digit to its right is less than 5, leave the identified digit as it is.
If the digit to its right is 5 or more, increase the identified digit by 1.
3. Drop all the digits to the right of the identified digit.

Round off 7.238 to the nearest **hundredth**

7.2[3]8
↓
7.2[3]8
7.2[4]8
7.24

7.238 ≈ 7.24

3.2 PRACTICE PROBLEMS

Round off each of the following decimals to the place indicated.

1. 9.376 to the nearest hundredth
2. 0.329 to the nearest tenth
3. 12.2685 to the nearest thousandth
4. 16.998 to the nearest hundredth

3.3 ADDING AND SUBTRACTING DECIMALS

To add or subtract decimals, first arrange the numbers vertically with the decimal points lined up, and then proceed as though the numbers were whole numbers (ignoring the points). After performing the operation, place a decimal point in the result directly below the other points.

When the decimals have a different number of decimal places, it is usually helpful to fill in zeros on their right. As discussed in the

4.79 + 5.83
↓
4.79
+5.83
─────
10.62

preceding section, these zeros will not change the values of the decimals.

For example,

$$17.32 - 9.246$$
$$\downarrow$$
$$\begin{array}{r} 17.320 \\ -9.246 \\ \hline 8.074 \end{array}$$

SAMPLE PROBLEM
Problem 9: Add 5.73 + 17 + 8.296.

Explanation and Procedure	Math Steps
1. Line up the decimal points and add. Remember that the whole number, 17, has an unwritten decimal point on its right. As indicated, the zeros written on the right are optional, but helpful.	5.73 5.730 17. or 17.000 + 8.296 + 8.296 ――― ――― 31.026 31.026

SAMPLE PROBLEM
Problem 10: Subtract 73 − .46.

Explanation and Procedure	Math Steps
1. Place a decimal point to the right of the whole number, 73, and line up the decimal points.	73. −.46 ――
2. Fill in two zeros and subtract. Note that for subtraction, the zeros written on the right are usually not optional.	73.00 −.46 ―― 72.54

3.3 PRACTICE PROBLEMS

Perform the following additions and subtractions.

1. 4.32 + .168 + 17
2. 13.642 − 4.19
3. 5.6 − 2.931
4. 56 − .32

3.4 MULTIPLYING DECIMALS

```
 2.13   (2 places)
× 3.4   (1 place)
 ―――
  852
 639
 ―――
7.242   (3 places)
```

To multiply decimals, again proceed as though the numbers were whole numbers. This time, however, instead of lining up the decimal points, line up the right-most digits (as in whole number multiplication). After multiplying, place a decimal point in the result so that the number of places to its right is equal to the sum of the decimal places in the two numbers just multiplied.

SAMPLE PROBLEM

Problem 11: Multiply $3.4 \times .007$.

Explanation and Procedure	Math Steps
1. Arrange the numbers as though they were whole numbers (with the right most digits lined up), and multiply them.	$\begin{array}{r} 3.4 \\ \times .007 \\ \hline 238 \end{array}$
2. Since the result should have 4 decimal places (1 + 3 = 4), fill in a zero in front of the 2 to act as a placeholder.	$\begin{array}{rl} 3.4 & \text{(1 place)} \\ \times .007 & \text{(3 places)} \\ \hline .0238 & \text{(4 places)} \end{array}$

When multiplying decimals by a *power of 10* (10, 100, 1000, etc.), we can use the following shortcut:

To Multiply a Decimal by a Power of 10:

1. Move the decimal point to the *right* as many places as the number of zeros in the power of 10.

Consider the examples below.

$$\begin{array}{r} 7.382 \\ \times 10 \\ \hline 73.820 \end{array} \qquad \begin{array}{r} 7.382 \\ \times 100 \\ \hline 738.200 \end{array}$$

$$73.82 \times 10 = 7\underset{\rightarrow}{.3}.82 = 73.82$$

$$73.82 \times 100 = 7\underset{\rightarrow}{.38}.2 = 738.2$$

As you can see, when we multiply by 10 (*one zero*), we move the decimal point *one place* to the *right*, and when we multiply by 100 (*two zeros*), we move the decimal point *two places* to the *right*.

3.4 PRACTICE PROBLEMS

Perform the following multiplications.

1. 5.14×2.7
2. $6.3 \times .002$
3. 5.347×100
4. 6.23×1000

3.5 DIVIDING A DECIMAL BY A WHOLE NUMBER

When dividing a decimal by a whole number, we first position a decimal point for the result directly above the point in the dividend. Then we proceed to divide in the usual manner, treating the dividend as a whole number.

$$\begin{array}{r} 3.12 \\ 4 \overline{\smash{)}12.48} \end{array}$$

Correct: $9\overline{)\,.216}^{\,.024}$

Incorrect: $9\overline{)\,.216}^{\,.24}$

As shown in the margin, sometimes it is be necessary to fill in zeros, to act as place-holders, between the decimal point and the first digit in the result. It is incorrect to leave these places blank.

If the division is not exact, we can either express the remainder as a fraction, or continue to divide by adding zeros to the right of the last digit in the dividend. Remember that when we add zeros to the right of a decimal, we do not change its value. Both procedures are shown below.

$$
\begin{array}{r}
5.8\tfrac{3}{4} \\
4\overline{)23.5} \\
20 \\
\hline
35 \\
32 \\
\hline
3
\end{array}
\qquad
\begin{array}{r}
5.875 \\
4\overline{)23.500} \\
20 \\
\hline
35 \\
32 \\
\hline
30 \\
28 \\
\hline
20 \\
20 \\
\hline
0
\end{array}
$$

Notice that by adding zeros on the right, the division eventually ends. Consequently, we call the result a **terminating decimal**. Sometimes, however, regardless of how many zeros are added, the division never ends. For example,

$$
\begin{array}{r}
2.466... \\
3\overline{)7.400...} \\
6 \\
\hline
14 \\
12 \\
\hline
20 \\
18 \\
\hline
20 \\
18 \\
\hline
2
\end{array}
$$

Since the digit 6 keeps repeating, we call the result a **non-terminating repeating decimal**. In cases like these we have the following options:

$2.4666... \approx 2.47$

1. Approximate the result by rounding it off to a given place.

$2.4666... = 2.4\overline{6}$

2. Indicate the repeating digit(s) by placing a bar over it (them).

$2.4666... = 2.46\tfrac{2}{3}$

3. Express the remainder as a fraction.

SAMPLE PROBLEM

Problem 12: Divide $11\overline{)4}$.

Explanation and Procedure	Math Steps
1. Place a decimal point after the whole number 4, and position another point directly above it for the result.	$11\overline{)4} = 11\overline{)4.}^{\,.}$

| 2. By adding zeros to the right of the decimal point, keep dividing until either the division terminates, or the digits in the result start to repeat. | ```
      .3636...
   11)4.0000
      3 3
      ───
       70
       66
       ──
        40
        33
        ──
        70
        66
        ──
         4
         ⋮
``` |
| 3. Since the digits 36 are repeating, indicate them with a bar. | $.3636 = .\overline{36}$ |

When dividing decimals by a *power of 10* (10, 100, 1000, etc.), we can use a shortcut similar to the one for multiplying decimals by a power of 10.

> **To Divide a Decimal by a Power of 10:**
>
> 1. Move the decimal point to the *left* as many places as the number of zeros in the power of 10.

Consider the examples below:

```
       4.27                    .427
   10)42.70              100)42.700
      40                      40 0
      ──                      ────
       27                      2 70
       20                      2 00
       ──                      ────
        70                       700
        70                       700
        ──                       ───
         0                         0
```

$42.7 \div 10 = 4.2.7 \div 10 = 4.27$ $42.7 \div 100 = .42.7 \div 100 = .427$

As you can see, when we divide by 10 (*one zero*), we move the decimal point *one place* to the *left*, and when we divide by 100 (*two zeros*), we move the decimal point *two places* to the *left*.

3.5 PRACTICE PROBLEMS

Perform the following divisions.

1. $4\overline{).312}$ 2. $8\overline{)2.5}$ 3. $33\overline{)8}$ 4. $100\overline{)5.42}$

3.6 DIVIDING A DECIMAL BY A DECIMAL

To divide a decimal by a decimal, we change the problem into an equivalent problem in which the divisor is a whole number. To do this, we move the decimal point in the divisor all the way to its right,
```

$$.2\overline{)6.84} = .2.\overline{)6.8.4}_{\rightarrow\quad\rightarrow} \quad \begin{array}{r}34.2\end{array}$$

$$.23\overline{)4.6} = .23.\overline{)4.60.}_{\rightarrow\quad\rightarrow} \quad \begin{array}{r}20.\end{array}$$

and then move the decimal point in the dividend the same number of places to its right. After the decimal points are moved (and the divisor is a whole number), we proceed as in the last section.

When the dividend has fewer decimal places than the divisor, we make up the difference by adding as many zeros as necessary to the right of the dividend.

SAMPLE PROBLEM

Problem 13: Divide $.4\overline{)19}$.

Explanation and Procedure	Math Steps
1. Place a decimal point and one zero after the whole number, 19, and then move both decimal points one place to the right.	$.4\overline{)19} = .4\overline{)19.0}$ $= .4.\overline{)19.0.}_{\rightarrow\quad\rightarrow}$
2. Place a decimal point directly above the decimal point in the dividend, and divide.	$\begin{array}{r} 47. \\ 4.\overline{)190.} \\ \underline{16} \\ 30 \\ \underline{28} \\ 2 \end{array}$
3. Add one zero after the decimal point in the dividend, and continue to divide.	$\begin{array}{r} 47.5 \\ 4.\overline{)190.0} \\ \underline{16} \\ 30 \\ \underline{28} \\ 20 \\ \underline{20} \\ 0 \end{array}$

SAMPLE PROBLEM

Problem 14: Divide $.34\overline{).9}$ and round off the result to the nearest tenth.

Explanation and Procedure	Math Steps
1. Add one zero after the last digit in the dividend, and then move both decimal points two places to the right.	$.34\overline{).9} = .34.\overline{).90.}_{\rightarrow\quad\rightarrow}$
2. Place a decimal point directly above the decimal point in the dividend, and divide.	$\begin{array}{r} 2. \\ 34\overline{)90.} \\ \underline{68} \\ 22 \end{array}$

3. To round off the result to the nearest tenth, we must determine the digit in the hundredths place (one place more than asked for). Therefore, add two more zeros after the decimal point, and continue to divide.	$$\begin{array}{r} 2.64 \\ 34\overline{)90.00} \\ 68 \\ \overline{22\ 0} \\ 20\ 4 \\ \overline{1\ 60} \\ 1\ 36 \\ \overline{24} \end{array}$$
4. Since the digit in the hundredths place is less than 5, drop it.	$2.64 \approx 2.6$

3.6 PRACTICE PROBLEMS

Perform the following divisions.

1. $.8\overline{)6.34}$ 2. $.12\overline{)4.5}$ 3. $.13\overline{)5.642}$ 4. $.32\overline{)68}$

Perform the following divisions and round off each answer to the nearest tenth:

5. $.9\overline{)8.7}$ 6. $.14\overline{)5.7}$ 7. $.42\overline{)51}$ 8. $.14\overline{)9}$

3.7 COMPARING DECIMALS

As you recall, one of the methods we can use to compare the values of fractions is to change the fractions to a common denominator, and then compare the resulting numerators.

In order to apply this method to decimal fractions, we can use the procedure below:

> ***To Compare Decimal Fractions:***
> 1. Add zeros to the right of the decimals so that all of them have the same number of decimal places (and thus the same denominator).
> 2. Compare the numbers to the right of the decimal points (the numerators of the decimal fractions).

For example, to compare the decimals .57, .413, and .6, we first add zeros to their right so that all of them are three-place decimals, having the same denominator, thousandths.

$.57 = .570$
$.413 = .413$
$.6 = .600$

We then compare their numerators by comparing the numbers to the right of the decimal points. Since 600 is the largest number, .6 is the largest decimal; since 413 is the smallest number, .413 is the smallest decimal.

3.7 PRACTICE PROBLEMS

For each of the following groups of decimals, find the smallest decimal.

1. .5, .468, .52
2. 8.6, 9.002, 8.59
3. .09, .765, .8
4. .64, 1.002, .9

3.8 CHANGING DECIMALS TO FRACTIONS

Fractions written in the standard form $\frac{a}{b}$ are called *common fractions*. By using the definition of a decimal fraction as a fraction whose denominator is a power of 10, we can change a decimal fraction into the form of a common fraction.

$$.8 = \frac{8}{10} = \frac{4}{5}$$

$$.62 = \frac{62}{100} = \frac{31}{50}$$

$$.035 = \frac{35}{1000} = \frac{7}{200}$$

To Change a Decimal into a Common Fraction:

1. Form a fraction whose numerator is the number to the right of the decimal point, and whose denominator is the power of 10 corresponding to the number of decimal places in the given decimal: If the decimal has one place, make the denominator 10; if the decimal has two places, make the denominator 100; and so on.
2. Reduce the resulting fraction to lowest terms.

SAMPLE PROBLEM

Problem 15: Change the decimal $.37\frac{1}{2}$ into a common fraction, reduced to lowest terms.

Explanation and Procedure	Math Steps
1. Form a fraction whose numerator is the number to the right of the decimal point, and whose denominator is 100. (Note: This is a two-place decimal; the fraction $\frac{1}{2}$ does not occupy a place.)	$.37\frac{1}{2} = \dfrac{37\frac{1}{2}}{100}$
2. To simplify the complex fraction, first change the mixed number in the numerator to an improper fraction, and then divide the numerator by the denominator.	$= \dfrac{\frac{75}{2}}{100}$ $= \dfrac{75}{2} \div 100$ $= \dfrac{\cancel{75}^{3}}{2} \times \dfrac{1}{\cancel{100}_{4}} = \dfrac{3}{8}$

3.8 PRACTICE PROBLEMS

Change the following decimals to common fractions, reduced to lowest terms:

1. .048
2. 12.68
3. $.8\frac{3}{4}$
4. $9.07\frac{1}{2}$

3.9 CHANGING FRACTIONS TO DECIMALS

By using the interpretation of a fraction as a quotient, we can change a common fraction into the form of a decimal.

$$\frac{a}{b} = b\overline{)a}$$

To Change a Common Fraction into a Decimal:

1. Place a decimal point and zeros to the right of the number in the numerator. The exact number of zeros depends upon the number of decimal places desired in the result.
2. Divide the denominator into the numerator.

$$\frac{3}{8} = 8\overline{)3.000}$$
$$\begin{array}{r}.375\\ \underline{24}\\ 60\\ \underline{56}\\ 40\\ \underline{40}\\ 0\end{array}$$

This procedure provides us with another method of comparing the values of common fractions. For example, to compare the fractions $\frac{5}{8}$, $\frac{3}{5}$, and $\frac{2}{3}$, we can divide the denominators into the numerators, and then compare the resulting decimals. Since $\frac{5}{8} = .625$, $\frac{3}{5} = .600$, and $\frac{2}{3} = .666...$, we conclude that $\frac{2}{3}$ is the largest fraction and $\frac{3}{5}$ is the smallest fraction.

$$\frac{5}{8} = 8\overline{)5.000} \quad .625$$
$$\frac{3}{5} = 5\overline{)3.000} \quad .600$$
$$\frac{2}{3} = 3\overline{)2.000...} \quad .666...$$

The decimal equivalents of several common fractions are given below. Since these fractions are used so frequently in computation, it is worthwhile to memorize them.

Common Fraction/Decimal Equivalents

			$\frac{1}{5} = .2$		$\frac{1}{8} = .125$
$\frac{1}{2} = .5$	$\frac{1}{3} = .333\frac{1}{3}$	$\frac{1}{4} = .25$	$\frac{2}{5} = .4$	$\frac{1}{6} = .16\frac{2}{3}$	$\frac{3}{8} = .375$
	$\frac{2}{3} = .666\frac{2}{3}$	$\frac{3}{4} = .75$	$\frac{3}{5} = .6$	$\frac{5}{6} = .83\frac{1}{3}$	$\frac{5}{8} = .625$
			$\frac{4}{5} = .8$		$\frac{7}{8} = .875$

3.9 PRACTICE PROBLEMS

Change the following common fractions to decimal fractions.

1. $\frac{12}{15}$
2. $9\frac{27}{72}$
3. $\frac{4}{18}$
4. $6\frac{12}{36}$

CHAPTER 4

4.1 PERCENTS

Fractions whose denominators are **100** are special fractions which are usually referred to as **percents** ("per hundred"). For example, $\frac{87}{100}$ can also be referred to as 87 percent. Instead of writing out the word "percent," we use the percent symbol, **%**, and write 87%. Some more examples are given below.

Fraction	Percent
$\frac{41}{100}$	41%
$\frac{9}{100}$	9%
$\frac{58\frac{1}{2}}{100}$	$58\frac{1}{2}$%

Although most of the percents we encounter in our daily lives are less than 100%, percents can be equal to, or even greater than 100%.

For example, if Mandy invested all of the money in her savings account into a state bond, we could say she invested 100% of her money into the bond. Similarly, if Emma's salary was 25% more than Celine's salary, we could say that Emma's salary was 125% of Celine's salary.

As shown in the margin, these percents, when written as fractions with a denominator of 100, have a numerator equal to or greater than 100.

$$100\% = \frac{100}{100}$$
$$125\% = \frac{125}{100}$$

4.2 CHANGING PERCENTS TO DECIMALS

When performing calculations with percents, it is often easier to work with their decimal equivalents. Since a percent is a fraction whose denominator is 100, we can change a percent into its decimal equivalent by dropping the % sign and dividing the percent by 100. Consider the examples that follow.

Percent ⟶ Fraction ⟶ Decimal

$$87\% = \frac{87}{100} = 100\overline{)87.00}^{\,.87}$$

$$3\% = \frac{3}{100} = 100\overline{)3.00}^{\,.03}$$

$$12.9\% = \frac{12.9}{100} = 100\overline{)12.900}^{\,.129}$$

In each of these examples, since we are dividing by 100, the decimal point in the resulting decimal moves two places to the left of its original position in the percent.

Percent ⟶ Decimal

$$87\% = .87.\% = .87$$
$$3\% = .03.\% = .03$$
$$12.9\% = .12.9\% = .129$$

To Change a Percent into a Decimal:

1. Move the decimal point in the percent *two places* to the *left* and *drop* the % symbol.

Sometimes we need to do just the opposite, that is, change a decimal into a percent. To do this we simply reverse the process.

To Change a Decimal into a Percent:

1. Move the decimal point in the decimal *two places* to the *right* and *add* a % symbol.

Some examples are given below.

Decimal ⟶ Percent

$$.93 = .93. = 93\%$$
$$.7 = .70. = 70\%$$
$$.641 = .64.1 = 64.1\%$$

Note that in both procedures above, we move the decimal point exactly two places. A simple way of remembering the direction in which to move the point (left or right) is to think of the letters of the alphabet.

ABC [D] EFGHIJKLMNO [P] QRSTUVWXYZ

In other words, if we change from a **Decimal** to a **Percent**, we move the decimal point two places to the *right* (**D → P**); if we change from a **Percent** to a **Decimal**, we move the decimal point two places to the *left* (**D ← P**).

SAMPLE PROBLEM

Problem 16: Change $9\frac{1}{4}\%$ into its decimal equivalent.

Explanation and Procedure	Math Steps
1. Change the fraction within the percent, $\frac{1}{4}$ to its decimal equivalent, .25.	$9\frac{1}{4}\% = 9.25\%$
2. Move the decimal point in the percent two places to the left (D ← P) and drop the % symbol.	$9\frac{1}{4}\% = .09.25\% = .0925$

SAMPLE PROBLEM

Problem 17: Change 1.8 into its percent equivalent.

Explanation and Procedure	Math Steps
1. Add a zero to the right of the 8.	1.8 = 1.80
2. Move the decimal point in the decimal two places to the right (D → P), and add a % symbol.	1.8 = 1.80. = 180.%

4.2 PRACTICE PROBLEMS

Change the following percents into their decimal equivalents.

1. 8.7%   2. .5%   3. $19\frac{1}{2}\%$ 4. $16.8\frac{1}{4}\%$

Change the following decimals into their percent equivalents:

5. .913   6. .079   7. 6.3   8. 12

4.3 CHANGING PERCENTS TO FRACTIONS

There are two methods we can use to change a percent into a fraction. In the first, we use the definition of a percent and form a fraction having a denominator of 100. In the second, we change the percent into a decimal (as in the previous section), and then change the decimal into a fraction.

To Change a Percent into a Fraction:

Method 1: Place the percent over a denominator of 100, and drop the % symbol. Reduce the resulting fraction to lowest terms.

Method 2: Change the percent into a decimal ($D \leftarrow P$), and then change the decimal into a fraction. Reduce the result to lowest terms.

$$28\% = \frac{28}{100} = \frac{7}{25}$$

$$28\% = .28. = \frac{7}{25}$$

To change a fraction into a percent, we simply reverse the process.

To Change a Fraction into a Percent:

Method 1: Multiply the fraction by 100% and simplify the result.

Method 2: Change the fraction into a decimal (by dividing the denominator into the numerator), and then change the decimal into a percent ($D \rightarrow P$).

$$\frac{13}{20} = \frac{13}{20} \times 100\% = 65\%$$

$$\frac{13}{20} = 20\overline{)13.00}^{.65} = 65.\%$$

In the graph below, we see the percent equivalents of several common fractions. Remember that fractions less than 1 (proper fractions) are equivalent to percents less than 100%, and that fractions equal to or greater than 1 (improper fractions) are equivalent to percents equal to or greater than 100%.

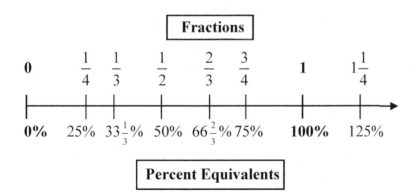

Page 31

SAMPLE PROBLEM

Problem 18: Change $12\frac{1}{2}\%$ into a fraction, reduced to lowest terms.

Explanation and Procedure	Math Steps
Method 1	
1. Place the percent over a denominator of 100, and drop the % symbol.	$12\frac{1}{2}\% = \dfrac{12\frac{1}{2}}{100}$
2. To simplify the complex fraction, first change the mixed number in the numerator, $12\frac{1}{2}$, into an improper fraction, and then divide the numerator by the denominator.	$= \dfrac{\frac{25}{2}}{100}$ $= \dfrac{25}{2} \div 100$ $= \dfrac{\cancel{25}}{2} \times \dfrac{1}{\cancel{100}_{4}}$ $= \dfrac{1}{8}$
Method 2	
1. Change the fraction $\frac{1}{2}$ in $12\frac{1}{2}\%$ to its decimal equivalent.	$12\frac{1}{2}\% = 12.5\%$
2. Move the decimal point two places to the left (D ← P), and drop the % sign.	$12.5\% = .12.5\% = .125$
3. Change the decimal into a fraction, and reduce the resulting fraction to lowest terms.	$.125 = \dfrac{125}{1000}$ $= \dfrac{125 \div 125}{1000 \div 125}$ $= \dfrac{1}{8}$

SAMPLE PROBLEM

Problem 19: Change 5.8% into a fraction, reduced to lowest terms.

Explanation and Procedure	Math Steps
Method 1	
1. Place the percent over a denominator of 100, and drop the % symbol.	$5.8\% = \dfrac{5.8}{100}$
2. To simplify the complex fraction, $\dfrac{5.8}{100}$ multiply the numerator and denominator by 10. This eliminates the decimal point in the numerator.	$= \dfrac{5.8 \times 10}{100 \times 10}$ $= \dfrac{58}{1000}$
3. Reduce the resulting fraction to lowest terms.	$= \dfrac{58 \div 2}{1000 \div 2}$ $= \dfrac{29}{500}$
Method 2	
1. Move the decimal point two places to the left ($D \leftarrow P$), and drop the % sign.	$5.8\% = .05.8\% = .058$
2. Change the decimal into a fraction, and reduce the resulting fraction to lowest terms.	$.058 = \dfrac{58}{1000}$ $= \dfrac{58 \div 2}{1000 \div 2}$ $= \dfrac{29}{500}$

SAMPLE PROBLEM

Problem 20: Change $\frac{11}{15}$ into a percent.

Explanation and Procedure	Math Steps
Method 1	
1. Multiply the fraction by 100% and simplify the result.	$\frac{11}{15} \times 100\% = \frac{11}{\cancel{15}} \times \frac{\cancel{100}^{20}\%}{1}$ $\phantom{\frac{11}{15} \times 100\%} {}_3$ $= \frac{220}{3}\%$ $= 73\frac{1}{3}\%$
Method 2	
1. Change the fraction to a decimal by dividing the denominator into the numerator. Stop dividing after two places, and reduce $\frac{5}{15}$ to $\frac{1}{3}$.	$\frac{11}{15} = 15\overline{)11.00} .73\frac{5}{15} = .73\frac{1}{3}$ $\phantom{\frac{11}{15} = 15)}\underline{10\ 5}$ $\phantom{\frac{11}{15} = 15)\ \ }50$ $\phantom{\frac{11}{15} = 15)\ \ }\underline{45}$ $\phantom{\frac{11}{15} = 15)\ \ \ \ }5$
2. Move the decimal point two places to the right (D → P), and add the % sign.	$\frac{11}{15} = .73.\frac{1}{3} \rightarrow 73\frac{1}{3}\%$

4.3 PRACTICE PROBLEMS

Change the following percents to fractions, reduced to lowest terms.

1. 16% 2. 125% 3. $8\frac{3}{4}\%$ 4. 12.8%

Change the following fractions to percents.

5. $\frac{9}{25}$ 6. $\frac{18}{30}$ 7. $\frac{35}{42}$ 8. $\frac{34}{51}$

ARITHMETIC REVIEW TEST

(Solutions to Arithmetic Review Test begin on Page 46)

1. What is 536,428 rounded off to the nearest 10,000?

   (A) 500,000
   (B) 530,000
   (C) 536,000
   (D) 537,000
   (E) 540,000

2. Which of the following is closest to $\dfrac{58 \times 3016}{97}$?

   (A) 1,700
   (B) 1,800
   (C) 1,900
   (D) 2,000
   (E) 2,100

3. The number of distinct prime divisors of 60 is

   (A) two
   (B) three
   (C) four
   (D) eight
   (E) ten

4. The fraction $\dfrac{450}{840}$ reduced to lowest terms is

   (A) $\dfrac{4}{7}$
   (B) $\dfrac{5}{8}$
   (C) $\dfrac{1}{2}$
   (D) $\dfrac{23}{42}$
   (E) $\dfrac{15}{28}$

5. $12\dfrac{7}{8} + 3\dfrac{5}{6} =$

   (A) $15\dfrac{35}{48}$
   (B) $15\dfrac{6}{7}$
   (C) $15\dfrac{17}{24}$
   (D) $16\dfrac{17}{24}$
   (E) $16\dfrac{41}{48}$

6. $6\dfrac{2}{5} - 2\dfrac{3}{4} =$

   (A) $3\dfrac{3}{20}$
   (B) $3\dfrac{13}{20}$
   (C) $3\dfrac{17}{20}$
   (D) $4\dfrac{7}{20}$
   (E) $4\dfrac{5}{9}$

7. $2\dfrac{1}{4} \times 8 \times \dfrac{2}{3} =$

   (A) $2\dfrac{2}{3}$
   (B) 12
   (C) $16\dfrac{1}{6}$
   (D) $16\dfrac{3}{7}$
   (E) $23\dfrac{1}{3}$

8. $6\frac{2}{3} \div 4 =$

   (A) $\frac{3}{80}$

   (B) $1\frac{1}{2}$

   (C) $1\frac{2}{3}$

   (D) $2\frac{1}{6}$

   (E) $6\frac{11}{12}$

9. Simplify $\frac{\frac{3}{4}+\frac{5}{7}}{2}$.

   (A) $\frac{4}{11}$

   (B) $\frac{41}{56}$

   (C) $1\frac{3}{14}$

   (D) $1\frac{5}{6}$

   (E) $2\frac{13}{14}$

10. Which of the following fractions is the largest?

    (A) $\frac{1}{2}$

    (B) $\frac{3}{4}$

    (C) $\frac{7}{9}$

    (D) $\frac{2}{3}$

    (E) $\frac{3}{5}$

11. If $\frac{N}{21} = \frac{40}{24}$, what is the value of N?

    (A) 3

    (B) 33

    (C) 35

    (D) 37

    (E) 43

12. What is 6.385 rounded off to the nearest tenth?

    (A) 6.38

    (B) 6.3

    (C) 6

    (D) 6.4

    (E) 6.39

13. $7.2 + 9 - 5.143 =$

    (A) 10.213

    (B) 10.687

    (C) 10.787

    (D) 10.813

    (E) 11.613

14. $8.4 \times .003 =$

    (A) .00252

    (B) .0252

    (C) .252

    (D) 2.52

    (E) 25.2

15. $.224 \div 7 =$

    (A) .0032

    (B) .03125

    (C) .032

    (D) .32

    (E) 31.25

16. $75 \div .08 =$

    (A) .9375
    (B) 9.375
    (C) 93.75
    (D) 937.5
    (E) 9375

17. Which of the following decimals is the smallest?

    (A) 1.003
    (B) .21
    (C) .1989
    (D) .5
    (E) .199

18. Which of the following is equivalent to 12.005?

    (A) $12\frac{1}{5}$
    (B) $12\frac{1}{4}$
    (C) $12\frac{1}{2}$
    (D) $12\frac{1}{20}$
    (E) $12\frac{1}{200}$

19. Which of the following is equivalent to $9.6\frac{3}{4}$?

    (A) $9\frac{27}{40}$
    (B) $9\frac{67}{100}$
    (C) $9\frac{2}{3}$
    (D) $9\frac{63}{100}$
    (E) $9\frac{63}{400}$

20. Which of the following is equivalent to $15\frac{3}{8}$?

    (A) 15.365
    (B) 15.37
    (C) 15.375
    (D) 15.38
    (E) 15.4

21. Which of the following is equivalent to $7\frac{1}{2}\%$?

    (A) .0075
    (B) .075
    (C) .75
    (D) 7.5
    (E) 7.5

22. Which of the following is equivalent to .003?

    (A) .03%
    (B) .3%
    (C) 3%
    (D) 30%
    (E) 300%

23. Which of the following is equivalent to 2.75?

    (A) 275%
    (B) 27.5%
    (C) 2.75%
    (D) .275%
    (E) .0275%

24. Which of the following is equivalent to $16\frac{1}{4}\%$?

   (A) $\frac{4}{25}$

   (B) $\frac{13}{80}$

   (C) $\frac{1}{25}$

   (D) $16\frac{1}{4}$

   (E) $\frac{17}{80}$

25. Which of the following is equivalent to $\frac{45}{72}$?

   (A) 62.5%

   (B) 16%

   (C) 6.25%

   (D) 1.6%

   (E) .625%

SOLUTIONS TO ARITHMETIC PRACTICE PROBLEMS

1.2 PRACTICE PROBLEMS

1. $3\boxed{4},682 \approx 35,000$

2. $5,\boxed{4}16,248 \approx 5,400,000$

3. $299,\boxed{9}61 \approx 300,000$

4. $\boxed{6}8,199 \approx 70,000$

1.4 PRACTICE PROBLEMS

1.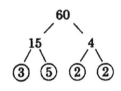

$60 = 3 \times 5 \times 2 \times 2$

2.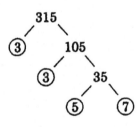

$315 = 3 \times 3 \times 5 \times 7$

3.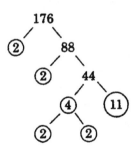

$176 = 2 \times 2 \times 2 \times 2 \times 11$

4.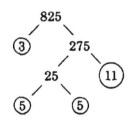

$825 = 3 \times 5 \times 5 \times 11$

2.2 PRACTICE PROBLEMS

1. $5\dfrac{3}{4} = \dfrac{4 \times 5 + 3}{4} = \dfrac{23}{4}$

2. $7\dfrac{3}{8} = \dfrac{8 \times 7 + 3}{8} = \dfrac{59}{8}$

3. $6\dfrac{2}{3} = \dfrac{3 \times 6 + 2}{3} = \dfrac{20}{3}$

4. $12\dfrac{1}{2} = \dfrac{2 \times 12 + 1}{2} = \dfrac{25}{3}$

5. $\dfrac{14}{3} = 3\overline{)14} = 4\dfrac{2}{3}$

6. $\dfrac{27}{2} = 2\overline{)27} = 13\dfrac{1}{2}$

7. $\dfrac{48}{4} = 4\overline{)48} = 12$

8. $\dfrac{6}{5} = 5\overline{)6} = 1\dfrac{1}{5}$

2.3 PRACTICE PROBLEMS

1. $\dfrac{3}{4} \begin{smallmatrix}5\\ \rightarrow \\ 5\end{smallmatrix} \dfrac{15}{20}$

2. $\dfrac{5}{9} \begin{smallmatrix}8\\ \rightarrow \\ 8\end{smallmatrix} \dfrac{40}{72}$

3. $\dfrac{4}{11} \begin{smallmatrix}6\\ \rightarrow \\ 6\end{smallmatrix} \dfrac{24}{66}$

4. $\dfrac{8}{3} \begin{smallmatrix}4\\ \rightarrow\\ \rightarrow\\ 4\end{smallmatrix} \dfrac{32}{12}$

2.4 PRACTICE PROBLEMS

1. $\dfrac{24 \div 2}{72 \div 2} = \dfrac{12 \div 2}{36 \div 2} = \dfrac{6 \div 2}{18 \div 2} = \dfrac{3 \div 3}{9 \div 3} = \dfrac{1}{3}$

2. $\dfrac{135 \div 3}{243 \div 3} = \dfrac{45 \div 3}{81 \div 3} = \dfrac{15 \div 3}{27 \div 3} = \dfrac{5}{9}$

3. $\dfrac{750 \div 10}{1250 \div 10} = \dfrac{75 \div 5}{125 \div 5} = \dfrac{15 \div 5}{25 \div 5} = \dfrac{3}{5}$

4. $\dfrac{420 \div 10}{560 \div 10} = \dfrac{42 \div 2}{56 \div 2} = \dfrac{21 \div 7}{28 \div 7} = \dfrac{3}{4}$

2.5 PRACTICE PROBLEMS

1. $\quad 5\dfrac{3}{4}$
   $+2\dfrac{3}{4}$
   $\overline{}$
   $\;\;7\dfrac{6}{4} = 8\dfrac{2}{4} = 8\dfrac{1}{2}$

2. $\quad 12\dfrac{5}{9}$
   $\;-7\dfrac{2}{9}$
   $\overline{}$
   $\;\;\;5\dfrac{3}{9} = 5\dfrac{1}{3}$

3. $\;\;{}^{7}\cancel{8}\dfrac{6}{6}$
   $-5\dfrac{5}{6}$
   $\overline{}$
   $\;\;2\dfrac{1}{6}$

4. $\;{}^{15}\cancel{16}\, 1\dfrac{4}{7} = 15\dfrac{11}{7}$
   $\qquad -9\dfrac{6}{7} = \;\;9\dfrac{6}{7}$
   $\qquad \overline{}$
   $\qquad\qquad\quad 6\dfrac{5}{7}$

2.6 PRACTICE PROBLEMS

1. $\;\;6\dfrac{3}{5} = 6\dfrac{21}{35}$
   $+2\dfrac{5}{7} = 2\dfrac{25}{35}$
   $\overline{}$
   $\;\;\;8\dfrac{46}{35} = 9\dfrac{11}{35}$

2. $\;\;2\dfrac{2}{3} = 2\dfrac{40}{60}$
   $\;\;4\dfrac{1}{4} = 4\dfrac{15}{60}$
   $+1\dfrac{3}{5} = 1\dfrac{36}{60}$
   $\overline{}$
   $\;\;\;7\dfrac{91}{60} = 8\dfrac{31}{60}$

3. $\;\;8\dfrac{2}{3} = 8\dfrac{4}{6}$
   $-3\dfrac{1}{2} = 3\dfrac{3}{6}$
   $\overline{}$
   $\;\;\;5\dfrac{1}{6}$

4. $19\dfrac{2}{5} = {}^{18}\cancel{19}\,1\dfrac{8}{20} = 18\dfrac{28}{20}$
   $-12\dfrac{3}{4} = \quad\;\; 12\dfrac{15}{20} = 12\dfrac{15}{20}$
   $\overline{}$
   $\qquad\qquad\qquad\qquad\; 6\dfrac{13}{20}$

2.7 PRACTICE PROBLEMS

1. $\dfrac{\cancel{2}^{1}}{\cancel{3}_{1}} \times \dfrac{\cancel{6}^{2}}{7} \times \dfrac{3}{\cancel{4}_{2}} = \dfrac{6}{14} = \dfrac{3}{7}$

2. $2\dfrac{2}{3} \times 1\dfrac{4}{5} = \dfrac{8}{\cancel{3}_{1}} \times \dfrac{\cancel{9}^{3}}{5} = \dfrac{24}{5} = 4\dfrac{4}{5}$

3. $5 \times \dfrac{2}{3} \times 1\dfrac{1}{5} = \dfrac{\cancel{5}^{1}}{1} \times \dfrac{2}{\cancel{3}_{1}} \times \dfrac{\cancel{6}^{2}}{\cancel{5}_{1}} = \dfrac{4}{1} = 4$

4. $1\frac{4}{5} \times 2\frac{2}{3} \times 2\frac{1}{2} = \frac{\overset{3}{\cancel{9}}}{\cancel{5}} \times \frac{\overset{4}{\cancel{8}}}{\cancel{3}} \times \frac{\overset{1}{\cancel{5}}}{\cancel{2}} =$

$\frac{12}{1} = 12$

2.8 PRACTICE PROBLEMS

1. $\frac{5}{6} \div \frac{4}{7} = \frac{5}{6} \times \frac{7}{4} = \frac{35}{24} = 1\frac{11}{24}$

2. $8 \div \frac{4}{9} = \frac{8}{1} \times \frac{9}{\cancel{4}} = \frac{18}{1} = 18$

3. $2\frac{2}{3} \div 1\frac{7}{9} = \frac{8}{3} \div \frac{16}{9} = \frac{\cancel{8}}{\cancel{3}} \times \frac{\cancel{9}}{\cancel{16}} =$

$\frac{3}{2} = 1\frac{1}{2}$

4. $2\frac{2}{5} \div 6 = \frac{12}{5} \div \frac{6}{1} = \frac{\cancel{12}}{5} \times \frac{1}{\cancel{6}} = \frac{2}{5}$

2.9 PRACTICE PROBLEMS

1. $\dfrac{\frac{2}{5}}{\frac{3}{4}} = \frac{2}{5} \div \frac{3}{4} = \frac{2}{5} \times \frac{4}{3} = \frac{8}{15}$

2. $\dfrac{4 - \frac{2}{7}}{6} = \dfrac{7 \times 4 - \cancel{7} \times \frac{2}{\cancel{7}}}{7 \times 6} = \frac{28 - 2}{42} =$

$\frac{26}{42} = \frac{13}{21}$

3. $\dfrac{\frac{1}{2} + \frac{2}{3}}{5} = \dfrac{\cancel{6} \times \frac{1}{\cancel{2}} + \cancel{6} \times \frac{2}{\cancel{3}}}{6 \times 5} = \frac{3+4}{30} = \frac{7}{30}$

4. $\dfrac{\frac{3}{4} + 2}{\frac{1}{2}} = \dfrac{\cancel{4} \times \frac{3}{\cancel{4}} + 4 \times 2}{\cancel{4} \times \frac{1}{\cancel{2}}} = \frac{3+8}{2} =$

$\frac{11}{2} = 5\frac{1}{2}$

2.10 PRACTICE PROBLEMS

1.

$\frac{5}{9}$ is the larger.

2.

$\frac{11}{15}$ is the larger.

3. Comparing $\frac{5}{7}$ and $\frac{2}{3}$:

$\frac{5}{7}$ is the larger.

Comparing $\frac{5}{7}$ and $\frac{3}{4}$:

$\frac{3}{4}$ is the largest.

2.11 PRACTICE PROBLEMS

1. $\dfrac{7}{9} = \dfrac{N}{72}$ (cross-multiply)

   $9 \times N = 7 \times 72$

   $9 \times N = 504$

   $N = \dfrac{504}{9} = 56$

2. $\dfrac{19}{12} = \dfrac{76}{N}$ (cross-multiply)

   $19 \times N = 12 \times 76$

   $19 \times N = 912$

   $N = \dfrac{912}{19} = 48$

3. $\dfrac{N}{16} = \dfrac{15}{20}$ (cross-multiply)

   $20 \times N = 15 \times 16$

   $20 \times N = 240$

   $N = \dfrac{240}{20} = 12$

4. $\dfrac{30}{N} = \dfrac{45}{27}$ (cross-multiply)

   $45 \times N = 27 \times 30$

   $45 \times N = 810$

   $N = \dfrac{810}{45} = 18$

3.2 PRACTICE PROBLEMS

1. 9.3[7]6 ≈ 9.38
2. 0.[3]29 ≈ 0.3
3. 12.26[8]5 ≈ 12.269
4. 16.9[9]8 ≈ 17.00

3.3 PRACTICE PROBLEMS

1.  ```
     4.320
      .168
   +17.000

    21.488
   ```

2.  ```
    13.642
    -4.190
    ------
     9.452
    ```

3.  ```
     5.600
    -2.931
    ------
     2.669
    ```

4.  ```
    47.00
     -.32
    -----
    46.68
    ```

3.4 PRACTICE PROBLEMS

1.  ```
      5.14
     ×2.7
     ----
     3598
     1028
    ------
    13.878
    ```

2.  ```
      6.3
     ×.002
     -----
     .0126
    ```

3. $5.347 \times 100 = 534.7$
4. $6.23 \times 1000 = 6230.$

3.5 PRACTICE PROBLEMS

1.  ```
         .078
       _____
      4).312
         28
         --
         32
         32
         --
          0
    ```

2.  ```
          .3125
        _____
       8)2.5000
         2 4
         ---
          10
           8
          --
          20
          16
          --
          40
          40
          --
           0
    ```

3.  ```
          .24
        _____
      33)8.00
         6 6
         ---
         1 40
         1 32
         ----
            8
    ```

Page 42
    ```

4. $100\overline{)5.42} = 5.42 \div 100 = .0542$

3.6 PRACTICE PROBLEMS

1.
```
         7.925
   .8.)6.3.400
    →    →
         5 6
         ───
          7 4
          7 2
          ───
            20
            16
            ──
             40
             40
             ──
              0
```

2.
```
         37.5
  .12.)4.50.0
   →    →
        3 6
        ───
         90
         84
         ──
          60
          60
          ──
           0
```

3.
```
         43.4
  .13.)5.64.2
   →    →
        5 2
        ───
         44
         39
         ──
          52
          52
          ──
           0
```

4.
```
          2 12.5
  .32.)68.00.0
   →    →
       64
       ──
        4 0
        3 2
        ───
          80
          64
          ──
          160
          160
          ───
            0
```

5.
```
              ↓
         9.⬜6 6 ≈ 9.7
   .9.)8.7.0 0
    →  →
        8 1
        ───
          60
          54
          ──
           60
           54
           ──
            6
```

6.
```
              ↓
        40.⬜7 1 ≈ 40.7
  .14.)5.70.0 0
   →    →
       5 6
       ───
         10
          0
         ──
         10 0
          9 8
         ────
            2 0
            1 4
            ───
              6
```

7.
```
                ↓
         1 21.⬜4 2 ≈ 121.4
  .42.)51.00.0 0
   →    →
        42
        ──
         9 0
         8 4
         ───
          60
          42
          ──
          180
          168
          ───
           12 0
            8 4
           ────
            3 6
```

8.
```
              ↓
        64.⬜2 8 ≈ 64.3
  .14.)9.00.0 0
   →    →
       8 4
       ───
         60
         56
         ──
          40
          28
          ──
          120
          112
          ───
            8
```

Page 43

3.7 PRACTICE PROBLEMS

1. .500
   .468 ← smallest
   .520

2. 8.600
   9.002
   8.590 ← smallest

3. .090 ← smallest
   .765
   .800

4. .640 ← smallest
   1.002
   .900

3.8 PRACTICE PROBLEMS

1. $.048 = \dfrac{48}{1000} = \dfrac{6}{125}$

2. $12.68 = 12\dfrac{68}{100} = 12\dfrac{17}{25}$

3. $.8\dfrac{3}{4} = \dfrac{8\dfrac{3}{4}}{10} = 8\dfrac{3}{4} \div 10 =$

   $\dfrac{35}{4} \times \dfrac{1}{10} = \dfrac{7}{8}$

4. $.07\dfrac{1}{2} = \dfrac{7\dfrac{1}{2}}{100} = \dfrac{15}{2} \div \dfrac{100}{1} =$

   $\dfrac{15}{2} \times \dfrac{1}{100} = \dfrac{3}{40}$

   Thus, $9.07\dfrac{1}{2} = 9\dfrac{3}{40}$

3.9 PRACTICE PROBLEMS

1. $\dfrac{12}{15} = 15\overline{)12.0}$ quotient .8
   $\underline{12\,0}$
   0

2. $\dfrac{27}{72} = 72\overline{)27.000}$ quotient .375
   $\underline{21\,6}$
   $5\,40$
   $\underline{5\,04}$
   360
   $\underline{360}$
   0

   Thus, $9\dfrac{27}{72} = 9.375$

3. $\dfrac{4}{18} = 18\overline{)4.00}$ quotient $.22\dfrac{4}{18} = .22\dfrac{2}{9}$
   $\underline{3\,6}$
   40
   $\underline{36}$
   4

4. $\dfrac{12}{36} = 36\overline{)12.00}$ quotient $.33\dfrac{12}{36} = .33\dfrac{1}{3}$
   $\underline{10\,8}$
   $1\,20$
   $\underline{1\,08}$
   12

   Thus, $6\dfrac{12}{36} = 6.33\dfrac{1}{3}$

4.2 PRACTICE PROBLEMS

1. $8.7\% = .08.7 = .087$

2. $.5\% = .00.5 = .005$

3. $19\dfrac{1}{2}\% = 19.5\% = .19.5 = .195$

4. $16.8\dfrac{1}{4}\% = 16.825\% = .16.825 = .16825$

5. $.913 = .91.3\% = 91.3\%$

6. $.079 = .07.9\% = 7.9\%$

7. $6.3 = 6.30.\% = 630.\%$

8. $12 = 12.00.\% = 1200.\%$

4.3 PRACTICE PROBLEMS

1. $16\% = \dfrac{16}{100} = \dfrac{4}{25}$

2. $125\% = \dfrac{125}{100} = 1\dfrac{25}{100} = 1\dfrac{1}{4}$

3. $8\dfrac{3}{4}\% = \dfrac{8\dfrac{3}{4}}{100} = 8\dfrac{3}{4} \div 100 =$

   $\dfrac{35}{4} \div \dfrac{100}{1} = \dfrac{35}{4} \times \dfrac{1}{100} =$

   $\dfrac{\overset{7}{\cancel{35}}}{4} \times \dfrac{1}{\underset{20}{\cancel{100}}} = \dfrac{35}{80}$

4. $12.8\% = \dfrac{12.8 \times 10}{100 \times 10} = \dfrac{128}{1000} = \dfrac{16}{125}$

5. $\dfrac{9}{25} = 25\overline{)9.00}$ with quotient $.36$
   $7\ 5$
   $1\ 50$
   $1\ 50$
   0

   Thus, $\dfrac{9}{25} = .36 = .36.\underset{\rightarrow}{\%} = 36.\%$

6. $\dfrac{18}{30} = 30\overline{)18.0}$ with quotient $.6$
   $18\ 0$
   0

   Thus, $\dfrac{18}{30} = .6 = .60.\underset{\rightarrow}{\%} = 60.\%$

7. $\dfrac{35}{42} = 42\overline{)35.00}$ with quotient $.83\tfrac{14}{42} = .83\tfrac{1}{3}$
   $33\ 6$
   $1\ 40$
   $1\ 26$
   14

   Thus, $\dfrac{35}{42} = .83\dfrac{1}{3} = .83.\underset{\rightarrow}{\dfrac{1}{3}\%} = 83\dfrac{1}{3}\%$

8. $\dfrac{34}{51} = 51\overline{)34.00}$ with quotient $.66\tfrac{34}{51} = .66\tfrac{2}{3}$
   $30\ 6$
   $3\ 40$
   $3\ 06$
   34

   Thus, $\dfrac{34}{51} = .66\dfrac{2}{3} = .66.\underset{\rightarrow}{\dfrac{2}{3}\%} = 66\dfrac{2}{3}\%$

SOLUTIONS TO ARITHMETIC REVIEW TEST

1. E	6. B	11. C	16. D	21. B
2. B	7. B	12. D	17. C	22. B
3. B	8. C	13. C	18. E	23. A
4. E	9. B	14. B	19. A	24. B
5. D	10. C	15. C	20. C	25. A

1. E $5\boxed{3}6,428 \approx 540,000$

2. B $\dfrac{58 \times 3016}{97} \approx \dfrac{60 \times 30\cancel{00}}{1\cancel{00}} \approx 1,800$

3. B

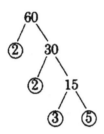

     The distinct prime divisors of 60 are 2, 3, and 5. Thus, there are three distinct prime divisors.

4. E $\dfrac{450}{840} = \dfrac{450 \div 10}{840 \div 10} = \dfrac{45 \div 3}{84 \div 3} = \dfrac{15}{28}$

5. D $\begin{aligned} 12\tfrac{7}{8} &= 12\tfrac{21}{24} \\ +3\tfrac{5}{6} &= 3\tfrac{20}{24} \\ \hline & 15\tfrac{41}{24} = 16\tfrac{17}{24} \end{aligned}$

6. B $\begin{aligned} 6\tfrac{2}{5} &= {}^5\cancel{6}\,1\tfrac{8}{20} = 5\tfrac{28}{20} \\ -2\tfrac{3}{4} &= 2\tfrac{15}{20} = 2\tfrac{15}{20} \\ \hline & 3\tfrac{13}{20} \end{aligned}$

7. B $2\tfrac{1}{4} \times 8 \times \tfrac{2}{3} = \dfrac{\cancel{9}^{3}}{\cancel{4}_{1}} \times \dfrac{\cancel{8}^{2}}{1} \times \dfrac{2}{\cancel{3}_{1}}$
   $= \dfrac{12}{1} = 12$

8. C $6\tfrac{2}{3} \div 4 = \dfrac{20}{3} \div \dfrac{4}{1} = \dfrac{\cancel{20}^{5}}{3} \times \dfrac{1}{\cancel{4}_{1}} =$
   $\dfrac{5}{3} = 1\tfrac{2}{3}$

9. B $\dfrac{\tfrac{3}{4} + \tfrac{5}{7}}{2} =$
   $\dfrac{\cancel{28}^{7} \times \tfrac{3}{\cancel{4}_1} + \cancel{28}^{4} \times \tfrac{5}{\cancel{7}_1}}{28 \times 2} = \dfrac{21 + 20}{56} = \dfrac{41}{56}$

10. C   Compare the fractions two at a time:

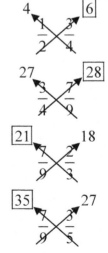

     Thus, $\dfrac{7}{9}$ is the largest.

Page 46

11. C

$$24 \times N = 21 \times 40$$
$$24 \times N = 840$$
$$N = \frac{840}{24} = 35$$

12. D $6.385 \approx 6.\boxed{3}85 \approx 6.4$

13. C
$$\begin{array}{r}7.2\\+9.0\\\hline 16.2\end{array} \qquad \begin{array}{r}16.200\\-5.413\\\hline 10.787\end{array}$$

14. B
$$\begin{array}{r}8.4\\\times.003\\\hline .0252\end{array}$$

15. C $.224 \div 7 = 7\overline{).224} = .032$

16. D $75 \div .08 = .08\overline{)75.00.0} = 937.5$

$$\begin{array}{r}72\\\hline 3\;0\\2\;4\\\hline 60\\56\\\hline 40\\40\\\hline 0\end{array}$$

17. C
$1.033 = 1.0330$
$.21 = .2100$
$.1989 = .1989 \leftarrow$ smallest
$.5 = .5000$
$.199 = .1990$

18. E $12.005 = 12\frac{5}{1000} = 12\frac{1}{200}$

19. A $.6\frac{3}{4} = \frac{6\frac{3}{4}}{10} = \frac{27}{4} \div \frac{10}{1} =$

$$\frac{27}{4} \times \frac{1}{10} = \frac{27}{40}$$

Thus, $9.6\frac{3}{4} = 9\frac{27}{40}$

20. C $\frac{3}{8} = 8\overline{)3.000} = .375$

$$\begin{array}{r}2\;4\\\hline 60\\56\\\hline 40\\40\\\hline 0\end{array}$$

Thus, $15\frac{3}{8} = 15.375$

21. B $7\frac{1}{2}\% = 7.5\% = .07.5 = .075$

22. B $.003 = .00.3\% = .3\%$

23. A $2.75 = 2.75.\% = 275\%$

24. B $16\frac{1}{4}\% = \frac{16\frac{1}{4}}{100} = \frac{65}{4} \div \frac{100}{1} =$

$$\frac{\cancel{65}^{13}}{4} \times \frac{1}{\cancel{100}_{20}} = \frac{13}{80}$$

25. A $\frac{45}{72} \approx \frac{45}{72} \times 100\% =$

$$\frac{4500\%}{72} = 62.5\%$$

APPENDIX

A.1 A TOUR OF THE BASIC CALCULATOR

Below is an example of a typical basic calculator. In addition to the usual keys of $+, -, \times, \div,$ and $=$, there are some special keys that can be very useful.

THE CLEAR KEYS

 Clear All

This key clears all pending results and resets the display to 0. On some basic calculators, the Clear All key clears the memory and on others it does not. In general, this key is used to clear everything and start over.

 Clear Entry

This key clears the number just entered on the keypad. It does not clear operations, like add or subtract. Even though a 0 shows in the display after pressing this key, all previous results are still pending. This is not a clear all key, but a key that clears only the number just entered.

THE MEMORY KEYS

Similar to a computer, the calculator has an internal storage location, called its *Memory*. The Memory is used to store a number, or the result of a sequence of operations for later use.

 Add to Memory

This key adds the number in the display to the number in the memory and stores the result in the memory. Initially the number in the memory is 0. After a number is added to the memory, an **M** appears somewhere in the display indicating that there is now a number in the calculator's memory.

Subtract From Memory

This key subtracts the number in the display from the number in the memory and stores the result in the memory.

▢ **Recall · Clear Memory**

This key has two functions. Pressing it once recalls the number in the memory to the display, while leaving a copy of the number in the memory. Pressing it twice recalls the number in the memory to the display and then clears the memory.

A.2 CALCULATOR KEYSTROKE NOTATION

To explore these keys in more detail, let's introduce a simple shorthand notation, called the *Calculator Keystroke Notation*, which we will use to indicate a sequence of keystrokes on the calculator. This notation is very similar to regular mathematical notation, with a few important differences. Below is a list of simple rules that the Calculator Keystroke notation will follow.

> *Using the Calculator Keystroke Sequence Notation*
> 1. Each keystroke sequence will begin with "KS:", which indicates that what follows is a sequence of calculator keystrokes to be pressed on the calculator, not a mathematical calculation.
> 2. All numbers are written as usual, as they are in a mathematical calculation.
> 3. All other keys are written in a small rectangle, representing the specific key on the calculator.
> 4. The number in the display at any point of the keystroke sequence is written above a small vertical arrow.

For example, to write the mathematical calculation $3 + 5 - 2 = 6$ in Calculator Keystroke Notation, we would write:

$$KS: 3\boxed{+}5\boxed{-}2\boxed{=}\overset{6}{\uparrow}$$

Note that the answer, 6, is not written after the equal sign, as in regular mathematical notation, but is written above a small arrow. When we see this arrow, we should understand that 6 is the answer showing in the display, and is not a key to be pressed.

A.3 THE CLEAR KEYS

There are two Clear Keys on the calculator: ▢, Clear All, and ▢, Clear Entry. These keys are used to either: (1) start the current calculation over; or (2) correct a numerical entry in the current calculation so that we don't have to start over.

For example, suppose we want to perform the calculation $3 + 5 - 2$, and by mistake we press the wrong number key, 7, instead of 5. We would like to correct this mistake without pressing the ▢ key and starting over.

To do this we would use the following Keystroke Sequence:

$$KS: 3\boxed{+}7\boxed{CE}\underset{\uparrow}{\overset{0}{}}5\boxed{-}\underset{\uparrow}{\overset{8}{}}2\boxed{=}\underset{\uparrow}{\overset{6}{}}$$

When the [CE] key is pressed directly after entering a number, the number is cleared and a new number can be entered. Remember that when we press the [CE] key, a 0 will appear in the display. This does not mean that we have cleared everything and must start over. It only means that the number 7 has been cleared, and that everything we have entered up to this point is still pending.

Since the [CE] only clears a numerical entry, let's look at how we can clear a mistaken operation ($+, -, \times, \div$) without starting over.

For example, suppose we want to perform the calculation $3 + 5 - 2$, and by mistake we press the wrong operation key, [×], instead of [−].

$$KS: 3\boxed{+}5\boxed{\times}\underset{\uparrow}{\overset{8}{}}$$

To correct this we would use the following Keystroke Sequence:

$$KS: 3\boxed{+}5\boxed{\times}\underset{\uparrow}{\overset{8}{}}\boxed{-}\underset{\uparrow}{\overset{8}{}}2=\underset{\uparrow}{\overset{6}{}}$$

In other words, when we press a different operation key directly after entering any operation key, the first operation is cleared and the second operation is the one that is performed.

In summary, to correct a mistake in the middle of a calculation, we can use one of the procedures below.

To Correct a Mistake in the Middle of a Calculation:

1. To correct a numerical entry, press the [CE] key immediately after entering the wrong number, and then enter the right number.
2. To correct an operation entry, press the right operation immediately after entering the wrong operation.
3. To clear all numbers and all operations, press the [C] key.

A.4 THE MEMORY KEYS

There are several computational situations in which using the Memory Keys can be quite helpful.

1. Using The Memory Keys to Perform Several Computations with the Same Number

By storing a number in the Memory, we won't have to enter the number each time we need to use it in a computation. Instead, we can enter the number into the Memory just once, and then recall it from the Memory each time we need it.

For example, suppose some state has a sales tax of 8.75% (.0875 as a decimal), and we want to compute the amount of sales tax on a shirt which costs $24.99. By multiplying the price of the shirt by the sales tax (as a decimal), we get 2.186625, or $2.19 rounded off to the nearest cent:

$$\text{KS}: 24.99 \; \boxed{\times} \; .0875 \; \boxed{=} \; \uparrow \quad 2.186625$$

This is a single computation and is straightforward. If, however, we wanted to compute the sales tax on each of several items, we would have to enter the number .0875 for each computation, leading to possible entry errors, and at the minimum, to press five keys just to enter the number. Imagine if the tax rate had even more digits.

Instead, we could enter the tax rate just once, store it in the calculator's memory, and recall it each time we need it for a computation.

For example, let us see how to use the Memory Keys to compute the amount of sales tax on each of three items: a shirt costing $24.99, a tie costing $17.49, and a jacket costing $74.99.

To store the sales tax of .0875 into memory, we would first take a quick look at the display to see if the memory is clear. If not, an **M** will appear in the display, and we would first clear the memory by pressing the [R·CM] key twice.

Then we would enter the number .0875 into the calculator so that it appears in the display. If we now press the Add to Memory Key, [M+], the number in the display, .0875, will be added to 0, the number in the calculator's memory, which is 0 when it is clear. The result of the addition, .0875 (.0875 + 0 = .0875), will be stored in the memory, replacing the 0 that was there.

$$\text{KS}: \underbrace{.0875 \; \boxed{M+}}_{} \; \uparrow \quad .0875$$

stores the tax rate in memory

To Store a Number into the Calculator's Memory

1. Check the display to see if the memory is clear. If an **M** appears, indicating it is not, press the [R·CM] key twice.
2. Enter the number into the calculator that you are going to use for several computations.
3. Press the [M+] key. That number will now be in the memory.

To continue and compute the amount of sales tax for the $24.99 shirt, we would enter the price of the shirt into the calculator. Notice that when we do this, the 24.99 replaces the .0875 in the display automatically, without having to first clear the .0875. With the price of the shirt,

$24.99, now in the display, we want to do what we did in the single computation–multiply the $24.99 by the sales tax, .0875. This time, however, instead of entering .0875 to use in the multiplication, we can recall it from the calculator's memory using just one key, [R·CM], the **Recall·Clear Memory key**. Remember that when we press this key once, the number in the memory is recalled (brought to the display), and a copy of the number remains in the memory, leaving it to be used for another calculation. As shown below, the sales tax on the shirt is the same result that we got before, $2.19, rounded off to the nearest cent.

$$\text{KS: } .0875 \underbrace{\boxed{M+} \uparrow}_{\text{stores the tax rate in memory}} \overset{.0875 \ 24.99}{24.99} \underbrace{\uparrow \boxed{\times} \boxed{R \cdot CM} \uparrow \boxed{=} \uparrow}_{\substack{\text{recalls the tax rate and multiplies} \\ \text{it by the price of the shirt}}}^{.0875 \ 2.186625}$$

Since the sales tax of .0875 remains in the calculator's memory until we clear it, we can continue computing the amount of sales tax on as many items as we wish simply by entering the price of the item into the calculator and multiplying it by the number in memory. We bring this number to the display by pressing the [R·CM] key once.

Therefore, now computing the amount of sales tax on the tie costing $17.49, we get

$$\text{KS: } .0875 \underbrace{\boxed{M+} \uparrow}_{\text{stores the tax rate in memory}} \overset{.0875 \ 24.99}{24.99} \underbrace{\uparrow \boxed{\times} \boxed{R \cdot CM} \uparrow \boxed{=} \uparrow}_{\substack{\text{recalls the tax rate and multiplies} \\ \text{it by the price of the shirt.}}}^{.0875 \ 2.186625} 17.49 \underbrace{\boxed{\times} \boxed{R \cdot CM} \uparrow \boxed{=} \uparrow}_{\substack{\text{recalls the tax rate and multiplies} \\ \text{it by the price of the tie.}}}^{.0875 \ 1.530375}$$

or $1.53, rounded off to the nearest cent.

And finally, computing the amount of sales tax on the jacket costing $74.99, we get

$$\text{KS: } .0875 \underbrace{\boxed{M+} \uparrow}_{\text{stores the tax rate in memory}} \overset{.0875 \ 24.99}{24.99} \underbrace{\uparrow \boxed{\times} \boxed{R \cdot CM} \uparrow \boxed{=} \uparrow}_{\substack{\text{recalls the tax rate and multiplies} \\ \text{it by the price of the shirt.}}}^{.0875 \ 2.186625} 17.49 \underbrace{\boxed{\times} \boxed{R \cdot CM} \uparrow \boxed{=} \uparrow}_{\substack{\text{recalls the tax rate and multiplies} \\ \text{it by the price of the tie.}}}^{.0875 \ 1.530375} 74.99 \underbrace{\boxed{\times} \boxed{R \cdot CM} \uparrow \boxed{=} \uparrow}_{\substack{\text{recalls the tax rate and multiplies} \\ \text{it by the price of the jacket}}}^{.0875 \ 6.561625}$$

or $6.56, rounded off to the nearest cent.

In summary, to perform several computations with the same number, we first store that number into the calculator's memory using the [M+] key, and then recall it to the calculator's display whenever we need it by pressing the [R·CM] once.

2. Using the Memory Keys to Keep a Running Total of a Sequence of Calculations

Another situation in which using the Memory Keys can be quite helpful is when we want to keep a running total of a sequence of computations. By adding the result of each individual computation to the number in the memory, the number in the memory will always be the current running total. For example,

Suppose that we are shopping at the local office supply store, and we want to keep a running total of our cost. In particular, let's say that we bought 5 boxes of envelopes at $8.99 per box, then 8 reams of paper at $7.49 per ream, and then 3 printer cartridges at 17.99 per cartridge.

First, we check to see that the memory is clear (no **M** in the display). Then, to compute the total cost of the envelopes, we multiply 5 boxes times $8.99 per box.

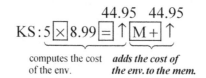

computes the cost
of the env.

Therefore, the total cost of the envelopes is $44.95. In order to start the running total, we would store this first total in the memory with the M+ key.

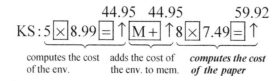

computes the cost   *adds the cost of*
of the env.   *the env. to the mem.*

Next, we compute the total cost of the paper.

$$KS: 5 \times 8.99 = \uparrow \underset{\text{44.95}}{M+} \uparrow 8 \times 7.49 = \underset{\text{59.92}}{\uparrow}$$

computes the cost   adds the cost of   *computes the cost*
of the env.   the env. to mem.   *of the paper*

Therefore, the total cost of the paper is $59.92. To keep the running total of the first two items, we again press the M+ key, which would add the number in the display, $59.92, to the number in the memory, $44.95, and then store the result in the memory (replacing the $44.95 that was there).

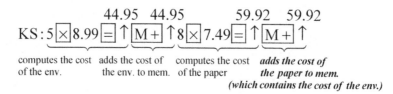

computes the cost   adds the cost of   computes the cost   *adds the cost of*
of the env.   the env. to mem.   of the paper   *the paper to mem.*
                                                         (which contains the cost of the env.)

At this point, notice that the number in the display is still 59.92, only the total cost of the paper. To actually see the running total of the envelopes and the paper, we would just press the R·CM key once to recall this total from the memory.

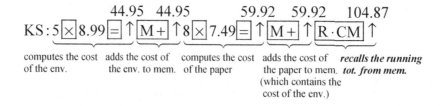

computes the cost   adds the cost of   computes the cost   adds the cost of   *recalls the running*
of the env.   the env. to mem.   of the paper   the paper to mem.   *tot. from mem.*
                                                            (which contains the
                                                            cost of the env.)

Thus, the running total of the first two items is $104.87.

Finally, we compute the total cost of the printer cartridges, add the result to the running total in the memory, and the recall this final sum from the memory.

$$\text{KS}: 5\boxed{\times}8.99\boxed{=}\underset{\underset{\text{of the env.}}{\text{computes the cost}}}{\uparrow}\overset{44.95}{\boxed{M+}}\underset{\underset{\text{the env. to mem.}}{\text{adds the cost of}}}{\uparrow}\overset{44.95}{}8\boxed{\times}7.49\boxed{=}\underset{\underset{\text{of the paper}}{\text{computes the cost}}}{\uparrow}\overset{59.92}{\boxed{M+}}\underset{\underset{\underset{\text{cost of the env.)}}{\text{(which contains the}}}{\text{the paper to mem.}}}{\underset{\text{adds the cost of}}{\uparrow}}\overset{59.92}{\boxed{R\cdot CM}}\underset{\underset{\text{from mem.}}{\text{running tot.}}}{\underset{\text{recalls the}}{\uparrow}}\overset{104.87}{}3\boxed{\times}17.99\boxed{=}\underset{\textit{\underset{of the cartridges}{computes the cost}}}{\uparrow}\overset{53.97}{\boxed{M+}}\underset{\underset{\underset{\text{of the env. + paper)}}{\text{has the tot. cost}}}{\underset{\text{to mem. (which}}{\text{of the cartridges}}}}{\underset{\text{adds the cost}}{}}\overset{158.84}{\boxed{R\cdot CM}}\underset{\underset{\text{from mem.}}{\text{running tot.}}}{\underset{\text{recalls the}}{\uparrow}}$$

In other words, the total cost of the printer cartridges is $53.97, and the total cost of all three items is $158.84.

In summary, to keep a running total of a sequence of calculations, we would perform each calculation, use the [M+] key to add the result to the number in the memory, and use the [R·CM] key to see the current running total whenever we want.

Printed in Great Britain
by Amazon